Medical Terminology

Master Your Medical Vocabulary by Learning to Pronounce, Understand and Memorize over 2000 of the Most Commonly Used Medical Terms.

Vincent Richards

in this book.

By reading this document, the reader agrees that under no circumstances is the author responsible for any losses, direct or indirect, which are incurred as a result of the use of information contained within this document, including, but not limited to, — errors, omissions, or inaccuracies.

Table Of Contents

Introduction

To master medical terminologies, you have to do more than just memorizing the terms, you need to explore the terms and how they can be created. This will reveal medical mysteries to you, and provide you with a better knowledge of every subject-based term.

Take this book as a private course that you must undertake to acquaint yourself with the medical terms used daily in hospitals, clinics, laboratories, pharmacies, and even health insurance companies.

The mastery starts with mastering the medical language itself. Hence, this book starts with the background of medical terminology; digging into the details of the formation of words, word parts, pronunciation, usage, and recognition. This will help you gain a better understanding of the art of creating and breaking down of words.

The second process involves understanding the different body systems and the words associated with them. Your knowledge of prefixes, suffixes, and root words will enable you to master various aspects of the terminology.

Please note that you are not obliged to read the whole of this book nor remember all the details in every chapter.

You can read the chapters randomly and skip any chapter that you don't need. You can always check the skipped sections another time.

There are some unique conventions in this book especially in making the terminologies easy to pronounce, that you may not find elsewhere. I deliberately used such a style in order to create a lively tone.

Systems of the body are used to divide the major chapters. Each system chapter explains how the body system works. The language with which I explained things is simple and easy to understand; so, you might see some informal words that you were not hitherto used to. This is done to neutralize the official terms that you are used to.

You can pick up this book at any time and study it in order to gain mastery of medical terminology. Understanding the how and why of medical terms is not less important than learning the terms themselves. This is the reason we put those chapters first. Nevertheless, feel free to study very hard the sections that you find very useful. Exude boldness and confidence while reading the book. Once you have mastered how these words are made, you will find it easy to memorize and master the terms in your daily life.

Chapter 1: Basic Concepts of Medical Terminology

Medical terminology refers to a unique vocabulary used by healthcare practitioners for efficient and effective communication. It includes various terms that describe the body organs, system, and functions (human anatomy and physiology). It also explains body locations, diagnostic imaging, diseases, laboratory testing, surgeries, diagnoses and lots more. All these deserve to have specific names, otherwise, medical professionals would find it difficult to communicate with one another. For instance, your doctor may cure you of a shoulder pain once you have complained to him. But while communicating with a surgeon, the doctor needs to be more specific.

Medical terminologies are uniform and consistent globally. This is because the words are originally derived from Greek and Latin words. Some of the terms may be long, they sometimes summarize a phrase to a single word. For example, *gastroduodenostomy* is a word that means "communication between the stomach and the first part of the small intestine."

Medical terminology is a vast field of knowledge. It is synonymous to learning the vocabulary of a foreign language. A a result of the ever-changing slangs in various fields, the medical vocabulary doesn't stop expanding. It seems to be an overwhelming task, however, there are methods that can be adopted in learning and mastering of words. Such methods can also help in guessing the meanings of new words.

Understanding Word Parts

Virtually every Medical term has its roots from the combination of prefixes, roots, and suffixes. Such component parts maintain their original meaning anywhere they are used. Learning these meanings would help you inanalyze and master many words. The root words are many because they fundamental to the meaning of every word. There are more roots than prefixes and suffixes combined.

Root

A root is the most important component of a word. Medical roots can signify a disease, procedure, or body part. While some roots can be found at the beginning of a word, others can be found after a prefix, before a suffix, or between a prefix and a suffix. Two or more roots can also be combined to form a word, as in cardi-o-vascular and cardi-o-pulmonary. O is the vowel that is most commonly used for such combinations.

Examples of roots used in different positions are listed below.

- **Angi**, a root which stands for the vessel, can be found in the word *angiodema*. This is an example of a root that appears at the beginning of a word.

- **Cephal**, a root word which stands for the head can be found in the word *encephalic*. This is an example of a root in the middle of a word.

- **Derm**, a root word which stands for skin, can be found in words such as 'scleroderma', an example of a root word used at the end of a word.

An example of the combination of roots is phototherapy. While photo means light, therapy means treatment.

Table 1: Common Exterior Root Words

Exterior Root	*What It Means*
Acr/o	Extremities
Amb/i	Both sides/double
Anter/o	Front
Aut/o	Self
Axill/o	Armpit, axilla

Blephar/o	Eyelid or eyelash
Brachi/o	Arm
Bucc/o	Cheek (on the face!)
Canth/o	Angle formed where eyelids meet
Capit/o	Head
Carp/o	Wrist
Caud/o	Tail/downward
Cephal/o	Head
Cervic/o	Neck or cervix (neck of the uterus)
Cheil/o, chil/o	Lip
Cheir/o, chir/o	Hand
Cili/o	Eyelash or eyelid, or small hair-like processes
Cor/e, cor/o	Pupil of eye
Crani/o	Skull
Cubit/o	Elbow
Dactyl/o	Fingers or toes
Derm/a, derm/o, dermat/o	Skin
Dors/i, dors/o	Back or posterior
Faci/o	Face
Gingiv/o	Gums in mouth
Gloss/o	Tongue
Gnath/o	Jaws
Inguin/o	Groin

Irid/o	Iris of eye
Labi/o	Lips
Lapar/o	Abdomen, loin, or flank
Later/o	Side
Lingu/o	Tongue
Mamm/a, mamm/o	Breast
Mast/o	Breast
Nas/o	Nose
Occipit/o	Back of the head
Ocul/o	Eye
Odont/o	Teeth
Omphal/o	Umbilicus
Onych/o	Nails
Ophthalm/o, ocul/o	Eyes
Optic/o, opt/o	Seeing, sight
Or/o	Mouth
Ot/o	Ear
Papill/o	Nipple
Pelv/o, pelv/i	Pelvis
Phall/o	Penis
Pil/o	Hair
Pod/o	Foot
Rhin/o	Nose
Somat/o	Body
Steth/o	Chest

Stomat/o	Mouth
Tal/o	Ankle
Tars/o	Foot
Thorac/o	Chest / thorax
Trachel/o	Neck or neck-like
Trich/o	Hair or hair-like
Ventr/i, ventr/o	Front of body

Table 2: Common Interior Root Words

Interior Root	*What It Means*
Abdomin/o	Abdomen
Acanth/o	Spiny or thorny
Acetabul/o	Acetabulum
Acromi/o	Acromium
Aden/o	Gland
Adip/o	Fat
Adren/o	Adrenal gland
Alveoli/o	Air sac
An/o	Anus
Angi/o	Vessel
Aort/o	Aorta
Arteri/o, arter/o	Artery
Arteriol/o	Arteriole
Aspir/o	To breathe in
Ather/o	Plaque, fat
Athr/o, articul/o	Joint
Atri/o	Atrium
Audi/o, aur/i	Hearing
Balan/o	Glans penis
Bio-	Life
Bronch/i, bronch/o	Bronchus
Bronchiol/o, bronchiol/i	Bronchiole
Carcin/o	Cancer
Cardi/o	Heart

Cellul/o	Cell
Cerebell/o	Cerebellum
Cerebr/i, cerebr/o	Cerebrum
Chol/e	Bile
Cholangi/o	Bile duct
Cholecyst/o	Gallbladder
Choledoch/o	Common bile duct
Chondr/i, chondr/o	Cartilage
Chrom/o, chromat/o	Color
Col/o, colon/o	Colon
Colp/o	Vagina
Cost/o	Rib
Cry/o	Cold
Crypt/o	Hidden
Cutane/o	Skin
Cyan/o	Blue
Cysti, cyst/o	Bladder or cyst
Cyt/o	Cell
Dipl/o	Double, twice
Duoden/o	Duodenum
Encephal/o	Brain
Enter/o	Intestine
Episi/o	Vulva
Erythr/o	Red
Esophag/o	Esophagus
Fibr/o	Fibers
Galact/o	Milk

Gastr/o	Stomach
Glyc/o	Sugar
Gynec/o	Female
Hemat/o, hem/o	Blood
Hepat/o, hepatic/o	Liver
Heter/o	Other, different
Hidr/o	Sweat
Hist/o, histi/o	Tissue
Hom/o, home/o	Same, alike
Hydr/o	Water, wet
Hyster/o	Uterus
Iatr/o	Treatment
Ile/o	Ileum (intestine)
Ili/o	Ilium (pelvic bone)
Intestin/o	Intestine
Jejun/o	Jejunum
Kerat/o	Cornea of eye, horny tissue
Lacrima	Tears
Laryng/o	Larynx
Leuk/o	White
Lipid/o	Fat
Lith/o	Stone (in gallbladder or kidney)
Lymph/o	Lymph vessels
Melan/o	Black
Men/o	Menses, menstruation

Mening/o	Meninges
Metr/a, metr/o	Uterus
My/o	Muscle
Myel/o	Bone marrow or spinal cord
Myring/o	Eardrum
Nat/o	Birth
Necr/o	Death
Nephr/o	Kidney
Neur/o	Nerve
Oophor/o	Ovary
Orchid/o, orchi/o	Testis
Oss/eo, oss/i, ost/e, ost/eo	Bone
Palat/o	Roof of mouth
Path/o	Disease
Peritone/o	Peritoneum
Pharmac/o	Drug
Pharyng/o	Pharynx
Phleb/o	Vein
Phren/o	Diaphragm
Pleur/o	Pleura, rib (side)
Pneum/a, pneum/o	Lungs
Pneum/ato, pneum/ono	Lungs
Poli/o	Gray matter of nervous system
Proct/o	Rectum, anus

Pulmon/o	Lungs
Py/o	Pus
Pyel/o	Pelvis of kidney
Rect/o	Rectum
Ren/i, ren/o	Kidney
Sacr/o	Sacrum
Salping/o	Fallopian tube
Sarc/o	Flesh
Scapul/o	Scapula
Sept/o	Infection
Splen/o	Spleen
Spondyl/o	Vertebra
Stern/o	Sternum
Tend/o, ten/o, tendin/o	Tendon
Testicul/o	Testis
Therm/o	Heat
Thorac/o	Chest
Thym/o	Thymus
Thyr/o	Thyroid gland
Thyroid/o	Thyroid gland
Tonsill/o	Tonsils
Trache/o	Trachea
Tympan/o	Eardrum
Ur/e, ur/ea, ur/eo, urin/o, ur/o	Urine
Ureter/o	Ureter
Urethr/o	Urethra

Uter/o	Uterus
Vas/o	Vas deferens
Vas/o, ven/o, ven/i	Vein
Vesic/o	Bladder
Viscer/o	Viscera (internal organs)
Xanth/o	Red, redness
Xer/o	Dry

Prefix

A prefix involves one or more letters attached to the beginning of a root. Most prefixes used in the medical field can also be seen in standard English vocabulary. To easily get the meaning of a word, you need to compare it with another word that begins with the same prefix. For instance, in the words 'antislavery' and 'antihistamine', 'anti' means against slavery and against histamine respectively. Histamine means the compound that produces allergic reactions.

Table 3: Common Prefixes

Prefix	*What It Means*
A-, an-	Lack of, without, not
Ante-	Before, in front of, or forward
Anti-	Opposing or against
Bi-	Double, two, twice, both
Co-, con-, com-	Together or with

De-	Down, or from
Di-	Twice or two
Extra-, extro-	Beyond, outside of, or outward
Hemi-, semi-	Half, half of
Hyp-, hypo-	Below, beneath, deficient
Hyper-	Above, excessive, beyond
Infra-	Below or beneath
Inter-	Between
Intra-	Within, inside
Intro-	Into, or within
Macro-	Large
Micro-, micr-	Tiny, small
Post-	After, or following, behind
Pre-, pro-	In front of, before, preceding
Retro-	Behind, backward
Semi-	Half
Trans-	Through or across
Tri-	Three
Ultra-	Excessive, beyond

Some prefixes share the same meaning, but they look different. Examples are listed below.

- Anti- and contra- stand for against.

- Dys and mal stand for bad or painful.

- Epi-, supra-, and hyper- stand for above.

- Endo- and intra- stand for within.

However, there are some prefixes whose meanings are opposite to each other despite looking or sounding similar. Some of them are:

- Ad- means toward; ab means away from (abduct).

- Post- means after; pre-, pro-, and ante- mean before.

- Hyper-, epi-, and supra-, mean above; sub-, intra-, and hypo mean below.

- Hypo- means deficient; hyper- means excessive.

- Brady- stands for slow; while tachy means fast.

Suffix

A Suffix consists of one or more letters attached to the end of a root. A suffix that starts with a consonant always has a combining vowel such as "o" placed before the suffix. In medical terminology, common suffixes include the addition of -y to a word to mean a procedure. An example is gastroscopy meaning the endoscopic examination of the stomach. The addition of -ly to a word means an act or process. An example is splenomegaly meaning the abnormal enlargement of spleen

Table 4: Common Suffixes

Suffix	*What It Means*
-ac, -al, -ar, -ary, -form, -ic, -ical, -ile, -oid, -ory, -ous, -tic	Related to, or pertaining to
-ate, -ize	Subject to, use
-ent, -er, -ist	Person, agent
-genic	Produced by
-gram	A written record
-graph	An instrument used to record
-graphy	Process of recording
-ia, -ism, -sis, -y	Condition or theory
-ian, -iatrics, -iatry, -ics, -ist, -logy	Medical specialties
-itis	Inflammation
-ologist	One who studies, specialist
-ology	Study of, the process of study
-oma	Tumor
-pathy	Disease, disease process
-phobia	Morbid fear of or intolerance
-scope	An instrument used to visually examine
-scopy	Process of visual examination

Word Derivations

As stated earlier, most medical terms have their roots traced from Greek and Latin. Sometimes, the original words and their meanings can be found in the text. For example, the word "coccyx" which means the tail end of the spine, has its origin from cuckoo due to the semblance between the cuckoo's bill and the spine end. The word "acrocyanosis" has its origin from "acr" meaning extremities and the "o" vowel which is joined with the root "cyan" meaning blue with -osis, a suffix which stands for the condition. The combination of all these means a situation of blue extremities.

Another example is the word "muscle" which originated from "mouse" because there is a semblance between the movement of muscles under the skin and the movement of a mouse.

With constant practice, you will become acquainted with various medical terms, and find it easy to interpret and combine them to form meaningful words.

Acronyms

An acronym is a word, or an abbreviation, formed using the first letters or syllables of other words. Uppercase letters are mostly used to express acronyms. There are many other common and uncommon acronyms in medical terminology. While some are identical, others have similar sounds with different meanings. Knowing the context in which they are used will go a long way in understanding a particular word or acronym.

Antonyms

An antonym is a word that means the opposite of another word. Examples are back and front, right and left, up and down, slow and fast, right and wrong. Some prefixes can also be paired as opposites as regards medical terms.

Table 5: Medical Antonyms

	Prefix	*What It Means*
1.	Ab-	Moving away from (abduction)
	Ad-	Drawing toward (adduction)
2.	Anterior-	Front
	Posterior-	Back
3.	Bio-	Life
	Necro-	Death
4.	Brady-	Slow
	Tachy-	Fast
5.	Cephalo-	Head (upward)
	Caudo-	Tail (downward)
6.	Endo-	Within, inside
	Exo-	Outside
7.	Eu-	Normal, well
	Dys-	Difficult, unwell

8.	Hyper-	Above or excessive
	Hypo-	Below or deficient
9.	Leuko-	White
	Melano-	Black
10	Pre-	Before or in front of
	Post-	After or behind
11	Proximal-	Near (think proximity)
	Distal-	Away from (think distance)
12	Superior-	Above
	Inferior-	Below

Eponyms

In the medical field, an eponym is a term that is named after the person that discovered a particular body part or disease. There are several tests and procedures named after the corresponding inventors.

Stated below are some examples of eponyms used for medical conditions.

- Alzheimer's disease is a kind of irreversible dementia Cushing's syndrome. The disease mostly occurs as a result of excess cortisol from the adrenal cortex.

- Stoke-Adams syndrome is a heart condition that causes loss of consciousness.

- Lyme disease is a multi-systematic disorder that is transmitted by ticks.

- Down syndrome, also known as trisomy 21, is a chromosomal disorder. It was formally called mongolism.

- Peyronie's disease is a deformity of the penis as a result of fibrous tissue found in the tunica albunigea.

- Addison's disease normally occurs as a result of insufficient production of hormones from the cortex of the adrenal gland.

- Parkinson's disease is a disease that causes weakness, tremors, and rigidity. It occurs as a result of the progressive degeneration of the nervous system.

There are some parts of the body that were named after the persons that discovered them. Some of them are:

- Bartholin's glands found in the female perineum.

- Wernicke's center which serves as the center of speech in the brain.

- Cowper's glands found below the male urethra.

- Ligament of Treitz located in the intestinal tract.

Medical procedures:

- Allen's test. This is a test done for occlusion of ulnar or radial arteries.

- Belsey Mark IV operation is a procedure used to correct gastroesophageal reflux.

- Heimlich maneuver is a method used for removing foreign objects from the airway of a choking victim.

Medical devices:

Some medical devices are named after their inventors. Examples are explained below.

- A Hickman catheter serves as a central venous catheter that is inserted to be used for the long term.

- The Foley catheter is an indwelling urinary catherer.

- A Malecot catheter is a tube used for gastronomy feedings.

Homonyms

A homonym is a word that has the same pronunciation but a different meaning with another word. Mostly, spellings of homonym words are different.

The word homonym is a Greek word, from homos, meaning same, and onyma, meaning name.

Examples of homonyms are feet and feat, pain and pane, bare and bear, peal and peel.

Table 5: Most Common Medical Homonyms

	Word	*What It Means*
1.	Cholic	An acid, related to bile
	Colic	Severe abdominal pain
2.	Humerus	A long bone in the upper arm
	Humorous	Funny
3.	Ileum	A portion of the colon
	Ilium	A part of the pelvic bone
4.	Jewel	A precious stone
	Joule	A unit of energy
5.	Lice	A parasite
	Lyse	To break
6.	Loop	An oval or circular ring, by bending
	Loupe	Magnifying glass or lens
7.	Mnemonic	To assist in remembering
	Pneumonic	Pertaining to the lungs (the "p" is silent)
8.	Mucous	The adjective form of mucus

		(resembling mucus)
	Mucus	Secretion of the mucous membranes
9.	Plain	Not fancy (plain x-rays)
	Plane	Anatomic (imaginary) level
10.	Pleural	Pertaining to the lung
	Plural	More than one
11.	Psychosis	Mental disorder
	Sycosis	Inflammation of hair follicles
12.	Radical	Extreme or drastic
	Radicle	A vessel's smallest branch
13.	Venous	Pertaining to a vein
	Venus	A planet

Soundalikes

	Word	*What It Means*
1.	Ablation	Surgical removal
	Oblation	A religious offering
2.	Access	A means of approaching
	Axis	Center
3.	Afferent	Towards the center
	Efferent	Away from the center
4.	Anecdote	A funny story
	Antidote	A remedy to treat poisoning
5.	Anuresis	Retention of urine in the bladder
	Enuresis	Bed-wetting

6.	Apparent	Clear, obvious
7.	Aberrant	Off course, abnormal
8.	Aural	Pertains to the ear
	Oral	Pertains to the mouth
9.	Callous	Hard like a callus, hardened thinking
	Callus	A hardened area of skin
10.	Carotid	Artery
	Parotid	Gland
11.	Cecal	Pertains to the cecum
	Fecal	Pertains to feces
12.	CNS	Central nervous system (abbreviation)
	C&S	Culture and sensitivity (a lab test)
13.	Discreet	Reserved or private
	Discrete	Separate
14.	Dysphagia	Difficulty eating or swallowing
	Dysphasia	Difficulty speaking
15.	Effusion	Escape of fluid into the tissue
	Infusion	To introduce fluid into vein or tissue
16.	Eczema	A type of dermatitis
	Exemia	Loss of fluid from blood vessels
17.	Ethanol	Alcohol
	Ethenyl	Vinyl
18.	Flanges	Projecting borders or edges
	Phalanges	Bones of the fingers or toes

19.	Graft	Tissue implant from one area to another
	Graph	Diagram
20.	Irradiate	To treat with radiation
	Radiate	To spread out from a center
21.	Joule	Energy
	Jowl	Flesh on the jaw
22.	Labial	Liplike
	Labile	Unstable
23.	Liver	Body organ
	Livor	Discoloration of skin after death
24.	Nucleide	A compound of nucleic acid
	Nuclide	A species of an atom
25.	Osteal	Bony
	Ostial	Pertaining to an ostium
26.	Palpation	To feel with the fingers
	Palpitation	Rapid heartbeat
27.	Perfusion	Pouring over or through
	Profusion	Abundant, much
	Protrusion	Jutting out
28.	Perineal	Referring to the perineum (genital area)
	Peritoneal	Referring to the peritoneum (in abdominal, pelvic cavities)
	Peroneal	A vein in the leg
29.	Pleuritis	Inflammation of the pleura of the lung

	Pruritus	Itching
30.	Precede	To come before
	Proceed	To carry on or continue
31.	Prostatic	Pertaining to the prostate gland
	Prosthetic	An artificial device replacing a body part
32.	Radical	Extreme; atoms in the uncombined state (free radicals)
	Radicle	A small branch of a vessel
33.	Scleroderma	Hardening of the skin
	Scleredema	Swelling of the face
34.	Shoddy	Poor quality of work
	Shotty	Resembles buckshot

Chapter 2: Body Structures

The body is composed of very large and complex structural units. These units consist of cells, tissues, organs, and body systems. These units work together to form the whole body.

The **body system** is a combination of body parts that work together to perform a related function. There are various methods used to describe the location of each of the body parts. They include body planes, body directions, body cavities, and structural units.

The Body Planes

These are imaginary horizontal and vertical lines used to divide the body into parts for ease of description.

The Horizontal Plane

This is a flat crosswise plane, such as looking at a horizon.

- A **transverse plane** is a type of horizontal plane that divides the body into upper (superior) and lower (inferior) portions. The transverse plane can be found at the waist or any level of the body.

The Vertical Planes

This is a plane located at the right angle of the horizon. It is characterized by an up-and-down movement.

- A **saggital plane** divides the body into left and right positions unequally.

- The **mid-sagittal plane** which divides the body into right and left in equal parts.

- A **frontal plane** divides the body into the front (anterior) and back (posterior) positions. It is also called **coronal plane**. It can be found at right angles to the sagittal plane.

Body Direction Terms

Pairs of contrasting body terms can be used to describe the relative location of sections of the body.

- **Ventr** means the belly side of the body; **-al** means pertaining to. **Ventral** (VEN-tral) thereby refers to the front or belly side of the body organ. The opposite of ventral is dorsal.

- **Dorsal** (DOR-sal) is the opposite of ventral. It originated from **dors**, which means the back of the organ or body; **-al** means pertaining to.

- **Anterior**, from **anter** meaning before or front, and **-ior** meaning pertaining to. Anterior means situated in the front. It also refers to the forward or front part of an organ. An example is the stomach, located in front of (anterior to) the pancreas. The opposite of anterior is posterior.

- **Posterior** (pos-TEER-ee-or) comes from **poster** meaning back and **-ior** meaning pertaining to. Posterior thereby means situated in the back. An example is the pancreas, located behind (posterior to) the stomach.

- **Superior** refers to the uppermost parts of the body. Superior means above, uppermost or toward the head. **Inferior** is the opposite of superior.

- **Inferior** refers to the lowermost part of the body. Inferior means below, lowermost or toward the feet.

- **Cephalic** (seh-FAL-ick) is from **cephal** which means head and **-ic** which means pertaining to. The opposite of cephalic is caudal.

- **Caudal** (KAW-dal) is from **caud** which means tail or lower part of the body, and **-al** which means pertaining to. Caudal means toward the lower part of the body. The opposite of caudal is cephalic.

- **Proximal** (PROCK-sih-mal) means situated very near to the beginning of the body structure or the midline. An example is the bone of the upper arm (humerus) whose proximal end forms part of the shoulder. The opposite of proximal is distal.

- **Distal**, the opposite of proximal, means situated very far from the beginning or midline of a body structure. An example is the distal end of the humerus which forms part of the elbow.

- **Lateral** refers to the direction nearer or toward the side of the body, away from the midline. An example is the nearness of the lateral ligament of the knee to the side of the leg. The opposite of lateral is medial. **Bilateral** means having or relating to two sides.

- **Medial** (MEE-dee-al), the opposite of lateral, refers to the direction toward or near the midline. An example is the nearness of the medial ligament of the knee to the inner surface of the leg.

Major Body Cavities

The ventral (front) and the dorsal (back) cavities are the two major body cavities which are spaces that contain and protect the internal organs of the body.

The Dorsal Cavity

This is located at the back, and tuns to the head. It houses the nervous system that controls the functions of the body. It is divided into:

- **The cranial cavity. Cranial** means pertaining to the skull. Therefore, the cranial cavity is housed in the skull, and it surrounds and protects the brain.

- **The spinal cavity.** This can be found in the spinal column. It surrounds and shields the spinal cord.

The Ventral Cavity

The ventral cavity is located at the frontal part of the body. It contains the body organs that sustain homeostasis. **Homeostasis** (hoh-mee-oh-STAY-sis) refers to the process by which the body maintains a constant internal environment. **Home/o** stands for constant, while **-stasis** stands for control.

The ventral cavity is divided into various portions. They are:

- The **thoracic** (thoh-RAS-ick) cavity, otherwise known as the thorax or chest cavity, protects and surrounds the lungs and the heart. The muscle that separates the abdominal and thoracic cavity is called the **diaphragm**.

- The **abdominal** (ab-DOM-ih-nal) cavity, simply known as the **abdomen** (AB-doh-men) contains the major organs of digestion.

- The **pelvic** (PEL-vick) cavity is the space within the hip bones. It contains the reproductive and excretory organs. There is no apparent division between the pelvic and the abdominal cavities.

- **Abdominopelvic** (ab-dom-ih-noh-PEL-vick) is from the word **abdomin/o** which means abdomen, while **pelv** means pelvis, and **-ic** means pertaining to. Abdominopelvic cavity means the two cavities as a single unit.

- **Inguinal** (ING-gwih-nal) means relating to the groin. It is used to refer to the lower area of the abdomen, including the groin, located at the upper part of the thigh.

Regions of the Thorax and Abdomen

These regions divide the abdomen and lower part of the thorax into nine different parts including the following:

- The right and left **hypochondriac** regions (high-poh-KON-dree-ack). The word originates from **hypo** which means below, **chondr/i** which means cartilage, and **-ac** which means pertaining to. The regions are the parts that cover the lower libs. Hypochondriac hereby means below the ribs.

- The **epigastric** region (ep-ih-GAS-trick). The term epigastric comes from **epi-** which means above, **gastr** which means stomach, and **-ic** which means pertaining to. The epigastric region is the part located above the stomach.

- The right and left **lumbar** regions. **Lumb** means lower back, while **-ar** means pertaining to. These regions are located close to the inward curve of the spine. Lumbar refers to the back area between the ribs and the pelvis.

- The **umbilical** region (um-BILL-ih-kal) is the region that surrounds the **umbilicus** (um-BILL-ih-kus), otherwise known as the navel or belly button. The abdominal wall has a pit in its center, where the umbilical cord was attached before birth.

- The right and left **iliac** (iLL-ee-ack) regions. **Ilii** means hip bone, while **-ac** means pertaining to. The right and left iliac regions are located on the hip bones.

- The **hypogastric** (high-poh-GAS-trick) region. **Hypo** means below; **gastr** means stomach; **-ic** means pertaining to. The hypogastric region can be found below the stomach.

Quadrants of the Abdomen

The abdomen can be divided into four imaginary parts. This makes it easy to describe the location of the abdominal organ. The Quadrants are:

- Left upper quadrant (LUQ)

- Right upper quadrant (RUQ)

- Left lower quadrant (LLQ)

- Right lower quadrant (RLQ)

The Peritoneum

The Peritoneum (pehr-ih-toh-NEE-um), is a membrane that holds and protects the organs within the abdominal cavity. A **membrane** is a thin layer of tissue which lines a cavity, covers a surface, or divides a space or organ.

- The **parietal peritoneum** (pah-RYE-eh-tal pehr-ih-toh-NEE-um) is the outer layer of the peritoneum which lines the inner part of the abdominal wall.

- The **mesentery** (MESS-en-terr-ee) is a fused double layer of the parietal peritoneum which attaches parts of the intestine to the interior abdominal wall.

- The **visceral peritoneum** (VIS-er-al pehr-ih-toh-NEE-um) refers to the inner layer of the peritoneum surrounding the organs of the abdominal cavity. **Visceral** means associated to the internal organs.

- **Retroperitoneal** (ret-roh-pehr-ih-toh-NEE-al) originates from **retro-**, which means behind, and **periton** means located beneath the peritoneum. An example is the kidney located retroperitoneal of the spinal column.

- **Peritonitis** (pehr-ih-toh-NIGH-tis) means inflammation of the peritoneum.

Chapter 3: Cell Structures

Cells are the simplest structural unit of the body. Tissues and organs are formed from a specialized collection of cells.

Cytology (sigh-TOL-oh-jee), 'cyt' - *cell*, '–ology' - *study*. Cytology is the study of the chemistry, physiology, anatomy, and pathology of cells.

Cytologists (sigh-TOL-oh-jist) are specialists in cell analysis and study.

The Structure of Cells

The **cell membrane** (MEM-brain) is the protective tissue which encloses and separates cell content from the outside environment.

The **nucleus** (NEW-klee-us) is found inside the cell, enclosed within the nuclear membrane. It has two vital functions: monitoring cell division and directing the functions of the cell.

The **Cytoplasm** (SIGH-toh-plazm) is the substances in the cell membrane, except the nucleus. 'cyto'- *cell*, 'plasm'- *cell substance*.

Stem Cells

Two features differentiate stem cells from other types of body cells:

- Other kinds of cells have specialized functions and die after a set period of time; whereas, stem cells are undifferentiated cells that are capable of self-regeneration.

- In certain instances, stem cells can become specialized cells. Examples are the cells that produce insulin in the pancreas and the cells in the heart muscle.

Adult Stem Cells

They are also referred to as **somatic stem cells**. They are unspecialized cells located among specialized cells in an organ or tissue. They are responsible for the repair and maintenance of the tissues where they are located. The presence of specialized structure or function is what makes a cell differentiated or undifferentiated.

Embryonic Stem Cells

This category of unspecialized cells is different from other specialized adult cells; they are able to transform into any adult cell type. They can multiply very fast in the lab and can be precursors for adult bone, muscle, blood, or liver cells.

Genetics

Genes are the basic unit of inheritance. They are responsible for hereditary diseases and physical features like eye color, skin, and hair.

Genetics ('gene' - *generation*; '–tics' - *related to*) is the study of how children (or offsprings) inherit genes from their parents and the significance of genes in disease and health. A **geneticist** (jeh-NET-ih-sist) is an expert in this discipline.

Dominant and Recessive Genes

Two copies of genes are transferred from parents to their offspring, one from each parent.

The inheritance of a dominant gene from one parent means that the child inherits that genetic feature or disorder. Freckles and Huntington's disease are inherited through a dominant gene.

The inheritance of recessive gene from both parents means that the disorder would be found in the child. For instance, sickle cell anemia is a hereditary red blood cell condition that is transferred by a recessive gene.

The inheritance of recessive genes from one parent and a normal gene from the other parent means that the disorder would be absent in the child; however, the child would have the sickle cell trait. Individuals with sickle cell traits can transfer the gene to their children.

The Human Genome

Genome (JEE-nohm) is the total set of genetic details of an individual. This genetic code was researched by the Human Genome Project and they discovered that there are 99% similarities among humans globally. The initial complete mapping of the human genome was finished after 13 years and was published in 2003.

Chromosomes

Every cell nucleus contains the genetic structure called **chromosomes** (KROH-moh-sohmes). Genes are contained in DNA, which makes up the chromosomes. Several genetic information is compactly stored in chromosomes and there are 100,000 genes in one chromosome.

- Any other cell apart from the sex cells (gametes) is called a **somatic cell**. **Somatic** means related to the body. There are 23 pairs of chromosomes in a somatic cell, 22 of these pairs are identical. The sex of a person depends on the last chromosome pair; usually, the last pair contains XX chromosomes in females and XY chromosome in males.

- The only kind of cell that does not have 46 chromosomes is the gamete, or sex cell (egg or sperm). Here, one egg contains 23 chromosomes. Usually, one of these is either an X or Y chromosome in males and an X chromosome in females. The fusion of the sperm and the eggs will produce an individual with 46 chromosomes, 23 from each parent. The sex of the offspring depends on the paternal X or Y chromosome.

- Chromosomal abnormalities can result in birth abnormalities. For instance, people with Down syndrome have 47 chromosomes.

DNA

The structure of the DNA of every living organism, found on the chromosome pairs, located in the nucleus, is the same. In humans, DNA is made of thousands of genes that are responsible for heredity, disease susceptibility, and physical features.

- DNA is an acronym for deoxyribonucleic acid and is in the nucleus of all cells, except red blood cells. It is arranged in chromosomes as two strands that are wrapped around each other, like spiral stairs, forming a double helix. Except in identical twins, DNA is unique to each

41

individual, and this is because one fertilized egg divides to form identical twins. Features such as fingerprints differentiate identical twins.

- In legal cases or genealogy research, individuals can be identified from tiny samples of DNA gotten from tissue or hair.

Genetic Mutation

An alteration in the sequence of a DNA molecule is called a genetic mutation. It is caused by environmental pollution and radiation exposure.

- An alteration in body cells is called a **somatic cell mutation**. The individual is affected by these changes, but they are not transmitted to the offspring.

- An alteration in the genes of the gamete is called a **gametic cell mutation**, and it can be transferred to offspring.

- The process of splicing or manipulation of genes for medical or scientific use is called **genetic engineering**. An example is the generation of human insulin from yeast.

Tissues

A collection of specialized cells that perform similar functions is called a tissue. **Histology** (hiss-TOL-oh-jee) 'hist'- tissue, '-ology'- study; Histology is the study of the structure and function of tissues. A **histologist** (hiss-TOL-oh-jist) is an expert that studies tissue organization ('-ologist'- expert). There are 4 major kinds of tissue: nerve tissue, epithelial tissues, muscle tissues, and connective tissues.

- **Nerve tissue**: is made of specialized cells that can conduct electrical impulses and react to stimuli.

- **Epithelial Tissues:** (ep-ih-THEE-lee-al) encloses all the external and internal surfaces of the body. They also form glands.

    o **Epithelium** (ep-ih-THEE-lee-um) is the differentiated epithelial tissue that forms the external layer of the mucous membrane and the skin epidermis.

    o **Endothelium** (en-doh-THEE-lee-um) is the differentiated epithelial tissue that forms the inner lining of organs, blood vessels, glands, lymph vessels, and body cavities.

- **Muscle Tissue:** is made of specialized cells that relax and contract.

- **Connective Tissues:** These tissues connect and support organs and other body tissues. There are 4 types:

  o **Liquid connective tissues** are the lymph and blood that transports waste and nutrients across the body.

  o **Dense connective tissues** like the cartilages and bones that makes the framework and joints of the body.

  o **Loose connective tissues** encloses organs, and support blood vessels and nerve cells.

  o **Adipose tissue** also called fat, creates a protective cushion and insulation.

Glands

A gland is a collection of epithelial cells that produce secretions. There are 2 main types.

- **Endocrine glands** (EN-doh-krin): 'Endo'- within, '–crine'- secrete. They are ductless glands that secrete hormones directly into the bloodstream and are transported to structures and organs across the body.

- **Exocrine glands** (ECK-soh-krin)": 'Exo'- outside, '-crine'- secrete. They secrete hormones into ducts that connect to other organs or outside the body, like sweat glands.

Chapter 4: The Skeletal System

The constituents of the skeletal system include ligaments, bones, joints, bone marrow, bursa, synovial membrane, synovial fluid, and cartilage.

- Red bone marrow is located in spongy bones and it produces blood cells.

- The joints, tendons, ligaments, and muscles facilitate a wide range of movements.

- Bones are the framework of the body. It supports and encloses internal organs.

The Formation of Bones

Initially, a child's skeleton is made of cartilage and delicate membranes; after 3 months, **ossification** (oss-us-fih-KAY-shun) takes place and it begins to harden into bone, this continues during adolescence.

Upon completion of growth, the process of formation continues as **osteoblasts** build new bones and **osteoclasts** destroy worn-out bones. The damage to the bone after regular activity and fractures is repaired by ossification.

The Structure of Bones

Bone is a type of connective tissue. The **enamel** (in the teeth) is the only tissue that is tougher than bone.

The Tissues of Bone

- **Periosteum** (pehr-ee-OSS-tee-um); 'peri'-enclosing, 'oste'- bone. It is the hard, fibrous tissue that covers the external surface of bones.

- **Compact bone** or **cortical bone** is the hard, strong bone that forms the protective external layer of bone.

- The **medullary cavity** (MED-you-lehr-ee) is the center cavity found within the shaft of long bones where it is enclosed by the compact bone. The yellow and red bone marrow are stored here.

- The **endosteum** (en-DOS-tee-um) forms the inner lining of the medullary cavity. 'endo'-inner, 'oste'- bone.

- **Spongy bone** or **cancellous bone** is lighter and has reduced strength, unlike compact bone. Usually, this bone is found in the inner and terminal parts of long bones like the femur. The red bone marrow is found in the spongy bone.

Bone Marrow

- **Yellow bone marrow** is located within the medullary of long bones. It stores fat.

- **Red bone marrow** is found in the spongy bone. It is a hemopoietic tissue that produces white blood cells, red blood cells, thrombocytes, and hemoglobin. **Hemopoietic** (hee-moh-poy-ET-ick) means the production of red blood cells. 'Hemo'- blood, '-poietic'-production.

Cartilage

Cartilage (KAR-tih-lidj) is the flexible, smooth, blue-white connective tissue that serves as a cushion between bones. It is found in the nose tip and outer ear.

- **Articular cartilage** (ar-TICK-you-lar KAR-tih-lidj) encloses the surface of bones where they form joints. Cartilage facilitates smooth movement of the joint and prevents friction between bones.

- The **meniscus** (meh-NIS-kus) is the curved fibrous cartilage located in joints like the knee.

Anatomic Landmarks of Bones

- The shaft of a long bone is the **diaphysis** (dye-AF-ih-sis).

- A **foramen** (foh-RAY-men) is a passage for nerves, blood vessels, and ligaments.

- The **epiphyses** (ep-PIF-ih-seez) are the broader end of long bones like the femur. Articular cartilages form a protective cover over each epiphysis. The **distal epiphysis** is the edge of the bone furthest from the midline of the bone. The **proximal epiphysis** is the bone edge closest to the midline.

- A **process** is a projection on the bone surface that creates an attachment for tendons and muscles.

Joints

Joints are the areas where two or more bones meet. They are grouped based on their composition, or the range of motion permitted.

Fibrous Joints

They contain rigid layers of dense connective tissues that firmly hold the bones together. In adults, they are called

sutures and do not permit any motion. In babies and children, a few fibrous joints permit movements before they solidify.

Fontanelles (fon-tah-NELLS) or soft spots are found on the skull of babies. Fontanelles are flexible and allow the infant to pass through the birth canal. They also facilitate skull growth in the first year. The sutures close once the child starts to mature, and the fontanelle slowly solidifies.

Cartilaginous Joints

Cartilaginous (kar-tih-LADJ-ih-nus) joints permit a limited range of movement and are made of bones that are totally connected by cartilage. They are found in the pubic symphysis where they facilitate movement during delivery, and where the ribs join the sternum. They also facilitate movement when breathing.

Synovial Joints

Synovial (sih-NOH-vee-al) joints are formed where two bones meet to allow a range of movements. They are classified based on the movements permitted:

- **Hinge joints** are found in the elbows and knees, permit unidirectional movements.

- **Ball-and-socket joints** are found in the

shoulders and hips permit a variety of motions.

Components of Synovial Joints

The constituents of synovial joints include

- **Synovial membrane** - produces synovial fluid and lines the capsule.

- **Synovial capsule** - the external layer of a strong, sleeve-like fibrous tissue that encloses the joint.

- **Synovial fluid** - courses through the synovial cavity, lubricates the joint and reduces friction.

- **Bursa** (BER-sah) - *Plural- bursae.* It is a fibrous sac that serves as a shock absorber to facilitate movement in areas that are prone to friction like the elbow, shoulder, and knee joints where a tendon moves over bone.

- **Ligaments** (LIG-ah-mentz) are fibrous bands that are formed when a bone meets another bone, or when a bone meets cartilage. The knee is a complex hinge joint that consists of a number of ligaments that allows a wide range of motion.

The Skeleton

There are 206 bones in the human adult skeleton. The number varies from 206-350 depending on the person's age. The skeleton is divided into appendicular and axial skeletons.

Appendicular skeleton: An appendage forms an attachment to a major part of the body. The appendicular skeleton permits the movement of the body and protects organs of reproduction, excretion, and digestion. The appendicular skeleton in humans have 126 bones that are grouped into the **upper extremities** (hands, wrists, forearms, arms, and shoulders) and the **lower extremities** (feet, ankles, legs, thighs, and hips). Extremities refer to the end of a body part like the leg or arm.

Axial skeleton: The main organs in the circulatory, respiratory, and nervous systems are protected by the axial skeleton. The human axial skeleton is made up of 80 bones of the head and body that are found in 5 places: the vertebral column, bones (ossicles) of the middle ear, the hyoid bone, rib cage, throat (between the hyoid and chin).

Bones of the Skull

The skull is made up of 14 bones of the face, 6 bones of the middle ear, and 8 bones of the cranium.

The Bones of the Cranium. The cranium (KRAY-nee-um); 'crani'- skull, '-um'- noun suffix. It forms a protective covering over the brain. The 8 bones of the Cranium are connected by rough fibrous joints called **sutures**.

- The **frontal bone** is the anterior part of the cranium that forms the forehead. It contains the frontal sinuses and forms the roof of the nose, eye sockets, and ethmoid sinuses.

- The **temporal bones** make up the base and sides of the cranium.

- The **parietal** (pah-RYE-eh-tal) bones are the biggest skull bones. The two parietal bones constitute the upper parts and roof of the cranium.

- The external **auditory meatus** (mee-AY-tus) is found in the temporal bone. It is formed by the opening of the external auditory canal of the ear. A **meatus** is the external passage of a canal.

- The **ethmoid** (ETH-moid) bone is a spongy, soft bone found on the sides and roof of the nose. It forms a part of each orbit (the socket that encloses and protects each eyeball) and a barrier between the brain and the nasal cavity.

- The **occipital** (ock-SIP-ih-tal) bone constitutes the base of the cranium and the posterior part of the skull.

- The **sphenoid** (SFEE-noid) bone is an asymmetrical, cuneiform bone found at the base of the skull. It connects to all the cranial bones. It forms part of the sides of the skull, the base of the cranium, and the sides and floor of the eye sockets.

The Auditory Ossicles (OSS-ih-kulz): These are 3 small bones called the malleus, incus, and stapes, that are found in each middle ear.

The Bones of the Face. The face consists of these 14 bones. A few of these bones contain sinuses which are cavities filled with air. They help to reduce the weight of the skull.

- Two **lacrimal** (LACK-rih-mal) bones form portions of the eye socket (orbit) at the inner angle.

- Two **zygomatic** (zye-goh-MAT-ick) bones, or **cheekbones**, joins the frontal bone to make up the forehead.

- Two **maxillary** (MACK-sih-ler-ee) bones, or **maxillae**, form majority of the upper jaw (singular-maxilla).

- The **mandible** (MAN-dih-bul), or the **jawbone**, is the only mobile bone in the skull. The TMJ, **temporomandibular joint**, connects the skull to the mandible.

- Two inferior **conchae** (KONG-kee or KONG-kay) are light, scroll-like bones that make up the inner part of the nose. **Concha** - singular.

- The **vomer** (VOH-mer) bone forms the base of the **nasal septum**, a cartilaginous wall separating the two nasal cavities.

- Two **palatine** (PAL-ah-tine) bones make up the floor of the nose and the anterior portion of the hard palate in the mouth.

Thoracic (thoh-RAS-ick) Cavity

Otherwise called the **rib cage**, is the bony structure that protects the lungs and heart. It is made up of the sternum, ribs, proximal part of the spinal column from the neck to the diaphragm, excluding the arms.

The Ribs. The twelve pairs of ribs or **costals** ('cost'- rib, 'al'- related), form a posterior attachment to the thoracic vertebrae. The **true ribs** are the first 7 pairs, and they form an anterior attachment to the sternum. The following 3 pairs are the **false ribs** and they are attached to the sternum anteriorly, through a cartilage. The final 2 pairs of ribs are **floating ribs** because they are not attached anteriorly to the vertebrae, but posteriorly.

The Sternum (STER-num). Also called the **breastbone**, this is a flattened, dagger-like bone found at the center of the chest. It forms the anterior part of the rib cage by combining with the ribs. There are 3 bony parts of the sternum:

- **The manubrium** (mah-NEW-bree-um) - the top of the sternum.

- **The body of the sternum**- forms the middle area.

- **The xiphoid process** (ZIF-oid) - the cartilaginous structure that makes up the lower part of the sternum.

The Shoulders

The **pectoral** (PECK-toh-rahl) or **shoulder girdle** that supports the arms and hands is formed during the development of the shoulders. A **girdle** is a structure that encloses the body. The bones of the shoulder are:

- The **scapula** (SKAP-you-lah), or shoulder blade. Scapulae - plural.

- The **clavicle** (KLAV-ih-kul) or collar bone, is a thin bone that attaches the scapula to the manubrium.

- The **acromion** (ah-KROH-mee-on) is a continuation of the scapula that forms the peak of the shoulders.

The Arms

- **Humerus** (HEW-mer-us) is the upper arm's bone. Humeri - plural.

- **Ulna** (ULL-nah) is the longest and bigger bone of the forearm. Ulnae - plural. The elbow joint is formed by the joining of the distal end of the humerus and the ulna.

- **Radius** (RAY-dee-us) is the smaller bone of the forearm. It extends upwards from the thumb.

- **Olecranon process** (oh-LEK-rah-non), or the funny bone is a big protrusion at the top of the scapula. It forms an obvious bulge on the elbow and contains a nerve that tickles if hit.

The Wrists, Hands, and Fingers

The eight **carpal** (KAR-pal) bones form the wrist. They create a narrow channel called the **carpal tunnel** which transmits the median nerve. The five **metacarpal** (met-ah-KAR-pal) bones form the palms. There are 14 **phalanges** (fah-LAN-jeez) in the fingers and toes. There are two thumb bones - the proximal and distal phalanges.

The Spinal Column

Also called the **vertebral column,** the spinal column supports the body and protects the spinal cord. A total of 26 **vertebrae** (VER-teh-bray) constitutes the spinal column (Vertebra-singular). Vertebral - relating to the vertebrae.

The Structures of Vertebrae

- The body of the vertebra is the anterior portion and it is designed to provide strength

- The **lamina** (LAM-ih-nah) is the posterior part of the vertebra. Laminae- plural. It extends to form the spinous and transverse processes which attach to tendons and muscles.

- The **vertebral foramen** is the central opening in the vertebra. It allows the passage of the spinal cord and protects it.

Intervertebral (in-ter-VER-teh-bral) Disks

It is composed of cartilages, and it acts as a cushion for

the vertebrae, separating them from one another. It also facilitates the movement of the vertebral column.

The Types of Vertebrae

- The **cervical vertebrae** (SER-vih-kal) are referred to as C1-C7. They are the first seven vertebrae that makes up the neck. 'Cervical' - related to the neck.

- The **thoracic vertebrae** (thoh-RASS-ick) are called T1-T12. They are the next 12 vertebrae. A pair of ribs are attached to each of these vertebrae and this forms the outward spine curvature. 'Thoracic'- related to the thoracic cavity.

- The **lumbar vertebrae** (LUM-bar) are called L1-L5. They are the third group of five vertebrae, and as a whole, they form the inward curvature of the lower spine. They bear the weight of the body and therefore, are the strongest and biggest vertebrae. 'Lumbar'- related to the back and sides from the ribs to the pelvis.

- The **sacrum** (SAY-krum) is the moderately curved triangular bone at the base of the spine that makes up the lower part of the back. There are 5 separate sacral bones at birth, but in children, it fuses to form one bone.

- The **coccyx** (KOCK-sicks), or tailbone makes up the terminal part of the spine, and is the fusion of 4 small vertebrae.

The Pelvic Girdle

The **pelvis** or **hips** make up the pelvic girdle, and it supports the lower extremities and protects internal organs. The pelvis is a ring of bone, shaped like a cup, and is found at the terminal part of the trunk. It comprises:

- The **ilium** (ILL-ee-um) is the flattened blade-shaped bone that makes up the sides and back of the pubic bone.

- The **ischium** (ISS-kee-um) makes up the lower end of the posterior part of the pubic bone. When sitting, it bears the weight of the body.

- The **sacroiliac** (say-kroh-ILL-ee-ack) is the slightly mobile joint between the posterior part of the ilium and the sacrum.

- The **pubis** (PEW-bis) makes up the anterior part of the pubic bone. It's under the urinary bladder.

- The ilium, pubis, and ischium are 3 distinct bones at birth. They start to fuse once the child begins to grow and they form the right and left pubic bones that are firmly held together by the pubic symphysis, a cartilaginous joint.

- The **acetabulum** (ass-eh-TAB-you-lum), or the **hip socket** is the big round cavity on both sides of the pelvis that joins the head of the femur to form the hip joint.

The Legs and Knees

The Femurs

The **femurs** (FEE-murz), or **thigh bones** are the biggest bones in the body. 'Femoral'- related to the femur. The acetabulum joins with the head of the femur. The narrow part beneath the head of femur is the **femoral neck**.

The Knees

- The **patella** (pah-TEL-ah), or **kneecap** is the bony anterior part of the knee.

- **Popliteal** (pop-LIT-ee-al) refers to the space posterior to the knee where muscles, ligaments, and vessels of the knee joint are found.

- The **cruciate** (KROO-shee-ayt) ligaments enable knee movements. The anterior and posterior cruciate ligaments are cross-shaped.

The Lower Legs

The fibula and tibia make up the lower leg.

- The **fibula** (FIB-you-lah) is the smaller bone of the lower leg.

- The **tibia** (TIB-ee-ah), or shinbone is the bigger bone of the lower legs that bears the weight.

The Ankles

- These are the joints that connect the foot and the lower leg, thus, facilitating important movements.

- There are seven short **tarsal** (TAHR-sal) bones in each ankle. They resemble the bones of the wrist but are bigger.

- The **malleolus** (mal-LEE-oh-lus) is a circular bony projection on the fibula and tibia. Malleoli-plural.

- The **calcaneus** (kal-KAY-nee-uss) or the **heel bone**, is the biggest tarsal bone.

- The **talus** (TAY-luss) is the bone of the ankle that joins the tibia and fibula.

The Feet and Toes

- The part of the foot attached to the toes is made up of the five **metatarsals** (met-ah-TAHR-salz)

- The bones of the toes are called **phalanges**. The biggest toe has two phalanges. Other toes have three phalanges

Chapter 5: The Muscular System

The muscles perform a number of functions in the body :

- They hold the body in an erect position, thereby aiding movement.

- About 85% of the heat generated by the body is due to muscular activities.

- Movement of food through the digestive system (peristalsis) is made possible by the muscles.

Structures of the Muscular System

The movement, stability, and shape of the body of all vertebrates depend on the body's synergistic interaction between the muscular and skeletal systems (which is usually jointly referred to as the **musculoskeletal system**).

The skeletal muscles are attached to the bones through the tendons and they are covered with **fascia**.

Muscle Fibers

Muscles are made up of long, slender cells known as **muscle fibers**. The muscle is made up of large amounts of fibers bound together by connective tissue (fascia).

The Fascia

The fascia is made up of a group of connective tissue that to envelop, separate, or bind muscles (or groups of muscles) together. The flexibility of the fascia makes muscular movement possible.

Myofascial (my-oh-FASH-ee-al) is a term commonly used in relation to muscle tissue and fascia.

My/o: Muscle, -fasci: fascia, -al: pertaining to.

Tendons

A tendon consists of a thin, fine band of dense, fibrous, non-elastic connective tissues. The primary role of the tendons is to attach muscles to bones, example in the the patellar tendon of the kneecap (patella).

Just as the fascia helps to connect muscles, the **aponeurosis**, a fine, sheet-like fibrous connective tissue, serves as a connective tissue between muscles and bones.

Types of Muscle Tissue

There are three different types of muscle tissue grouped according to their individual appearance and function.

They are:

- Skeletal muscles

- Smooth muscles and

- Myocardial

Skeletal Muscles (Striated or voluntary muscles)

As the name suggests, the skeletal muscles are found in connection to the skeletal system.

- **Based on their roles:** The skeletal muscles are known to facilitate voluntary body movement.

- **Based on appearance:** When viewed under a microscope, skeletal muscles can be differentiated from the others by the presence of striation.

Smooth Muscles (Visceral or Involuntary muscles)

These muscles have no striations and are found lining

the walls of visceral organs, valves, ducts, and vessels of the various organ systems. They mediate the various involuntary actions of the body and are controlled by the autonomic nervous system.

Myocardial Muscle (Cardiac muscle)

Myocardial (my-oh-KAR-dee-al) muscles or **cardiac muscle** is striated structurally just like the skeletal muscles, but functions involuntarily like smooth muscles

My/o: muscle, -cardi: heart, and -al: pertaining to.

The pumping action of the heart is due to the involuntary but constant contraction and relaxation of the myocardial muscle which we perceive as the heartbeat.

Muscular Contraction and Relaxation

The movement of our muscles is facilitated by the interaction of different muscles and innervation. Some muscles are organized in a way that movement is made possible by the antagonistic action of muscle pairs.

Muscle Innervation

Impulses are conducted from the central nervous system (CNS) to the various muscles by means of nerves. When a muscle is stimulated by a motor nerve, it brings about the contraction of such muscle. Damage to the nerve, either due to disease or traumatic injury, could disrupt the transduction of impulses to the muscles innervated by such damaged, hence, paralysis occurs on that part of the body.

The relationship between nerves and muscles is known as the **neuromuscular system**.

Neur/o: nerve, -muscul: muscle, -ar means pertaining to.

Antagonistic Muscle Pairs

Muscles act as antagonistic pairs. This means that, while one group of muscle contracts, the other relaxes and vice versa. When the muscle contracts, it becomes thicker and shorter, thus, the surface area at the center becomes larger.

Terms for Muscular Motions

The various movements of the muscle pairs can be described using various terms as discussed below.

Abduction and Adduction

Anatomically, the body is divided into two equal halves by an imaginary midline. Movement of the arm or leg away from this midline is known as **abduction** and all the muscles involved in such movement are known as the **abductor muscles**.

- Ab: away from, -duct: to lead, and -ion means action.

Adduction, on the other hand, involves the movement of the limbs towards the midline of the body, and muscles responsible for such movements are the **adductor muscles**.

- Ad: toward, -duct: to lead, and -ion: action.

Flexion and Extension

The term **flexion** simply means to bend a joint by decreasing the angle between two bones. For example, the bones of the knee and elbow are bent during flexion due to muscular contraction. Muscles that bring about such actions are known as **flexor muscles**.

- Flex: to bend, and -ion: action.

As the name suggests, during **extension**, the angle between the bones of a joint increase and the limb straightens out. **Extensor muscle** are responsible for the angular extension.

- Ex: away from, -tens: to stretch out, and -ion: action.

Hyperextension is the term used to describe the over-extension of any body part beyond its normal limit.

Elevation and Depression

The act of raising any part of the body is known as **elevation,** and muscles involved in raising body parts are known as **elevator muscles**.

Depression involves lowering of any body part below its normal limit, and the muscles involved in this act are called **depressor muscles**.

Rotation and Circumduction

Some bones can be rotated around an axis in a circular motion in an act called **rotation.** An example is the rotation of the shoulder joint. This movement is mediated by the **rotator muscle**.

Circumduction (ser-kum-DUCK-shun) on the other hand is the swinging motion of the far end of the limbs such as the legs and arms.

Pronation and Supination

Pronation (proh-NAY-shun) involves the rotation of the limbs to face backwards or downwards while **supination** (soo-pih-NAY-shun) is the act of rotating the limbs to face forward or upwards.

Dorsiflexion and Plantar Flexion

The bending of the toes of the feet upwards to face the

ankle is known as **dorsiflexion** (dor-sih-FLECK-shun). It decreases the angle between the tip of the toes and the ankle.

Plantar flexion (PLAN-tar FLECK-shun) on the other is the downward movement of the toes away from the ankle. It increases the angle between the tip of the toes and the ankle.

How Muscles are Named

Based on Their Origin and Insertion:

- All skeletal muscles have a point of origin and an insertion.

- The origin is where the muscle starts from. It is found near the body midline, and is usually rigid with little or no movements at all.

- The insertion is the point where the end of the muscle is attached to a bone or tendon. It is a movable attachment, farthest from the body midline.

- Some muscles have multiple points of origins such as the sternocleidomastoid muscle which has two points of origin - the sternum, clavicle, and the mastiod muscles.

Based on Their Action:

- As discussed earlier, some muscles perform a couple of functions and thus are named as such. For instance, the flexor carpi muscles that cause flexion of the toes.

Based on Their Location:

- Muscles can also be named based on their location on the body. For example, the orofacial muscles of the face and the pectoralis muscles (major and minor) of the anterior chest.

- The lateralis: toward the edges of the body, away from the midline.

- The medialis: toward the midline.

Based on the Fiber Direction:

Another criterion for naming muscles is based on the direction of their fibers. This includes the following:

- **Oblique muscles:** Slants or bends to a certain angle.

- **Transverse muscles:** Muscles that align horizontally to the body axis. An example is the transverse abdominis muscle.

- **Rectus Muscles:** Aligns vertically with the body. A good example is the rectus abdominis.

- **Sphincter:** These ring-like muscles that can either hinder or allow passage of substances through it. For instance, the pyloric sphincter.

Based on the Number of Divisions

- Muscles can also be named based on the number of divisions or muscles forming them such as the biceps brachii.

- Bi: two, and -ceps: head.

- Another similar example is the triceps brachii muscle, which is formed from three muscular divisions.

- Tri: three, and -ceps: head.

Based on Shape or Size

- Muscles such as the gluteus maximus are so named because of their broad size.

- The deltoid muscle of the shoulder is named as such because the shape resembles the Greek letter delta (more like an inverted triangle).

Based on Strange Reasons

The hamstrings muscle located at the back of the upper leg, for instance, is so named because butchers hang slaughtered pigs with these muscles.

The hamstring consists of three units of muscles:

- The biceps femoris

- Semi-tendinosus, and

- Semi-membranosus muscles.

The hamstring performs the antagonistic function of flexing the knee while the hip extends and vice versa.

Muscles and Their Functions

There are over 600 muscles in the human body, and each performs a specific function, some of which will be discussed briefly in this section.

Muscles of the Head:

- The muscle of the forehead is known as the **frontalis** (fron-TAY-lis) or **occipitofrontalis**.

  Function: It raises the eyebrows

- Muscles of the lower Jaw known as the **temporalis** (tem-poh-RAY-lis) muscle.

Function: It closes the mouth by moving the lower jaw up and back.

- One of the strongest muscles in the body is the **masseter** (mah-SEE-ter) muscle of the mouth.

  Function: It closes the mouth when chewing, by pulling the lower jaw up, thereby preventing food from dropping from the mouth while chewing.

Muscles of the Trunk

- The **pectoralis major** forms a larger percentage of the chest muscles in males but in females, the breast lay over this muscle.

- Located around the abdominal region are the **internal oblique** and **external oblique muscles**.

  Functions: The internal oblique muscles aids breathing, flex and rotate the spine, as well as provide support the abdominal organs.

  The external oblique muscle compresses the abdomen, flex and rotates the vertebral column, and they also flex the torso.

- Around the abdomen is the **rectus abdominis** (ab-DOM-ih-nus) muscle.

Function: Aids breathing, provides support to the spine, and helps to flex the trunk.

- Located on each side of the abdominal wall is the **transverse abdominis** muscle.

Function: Serves as a shock absorber and it becomes activated when an individual cough or laugh.

Muscles of the Shoulders and Arms

- **The deltoid muscle**: Serves as a muscular covering for the shoulder.

- **The trapezius muscle:** Aids the movement of the head and the scapula (shoulder blade).

- **The biceps brachii**: Found on the anterior part of the upper arm. It helps to flex the elbow.

- **The Triceps Brachii:** Found on the posterior part of the upper arm. It helps to extends the elbow.

Muscles of the Legs

- Extending from the hip to the knee is the rectus femoris (FEM-or-iss).

- Four different muscles make up the quadriceps

femoris, including the vastus medialis and the vastus lateralis involved in the flexion and extension of the leg at the knee.

- The hamstring are involved in the extension of the hip extension and flexion of the knee.

- The gastrocnemius muscle (the calf muscle). The name is coined from a latin word which means the stomach of the leg because of the way the muscles bulges out. It bends the foot downwards and flexes the knee.

Chapter 6: Nervous System

The control of all the activities in the body is handled by the nervous system that has the brain as it's core organ. The body dies as soon as the brain stops working. The nervous system is made of other key organs including the sensory organs and nerves.

Divisions of the Nervous System

The nervous system has two main subdivisions: the peripheral nervous system, and the central nervous system.

- The components of the peripheral nervous system are 31 pairs of peripheral nerves branching outwards from the spinal cord, and 12 pairs of cranial nerve branching outwards from the brain. The peripheral nervous system transmits impulses from and to the nervous system.

- The central nervous system is made up of the spinal cord and brain. It controls other functions in the body and also processes received information.

The Nerves

A **nerve** refers to a single neuron or a collection of neurons that serves as a connection between the spinal cord, brain, and other body parts. A **tract** is a collection of nerve fibers in the spinal cord or brain.

- Transmission of nerve signals to the brain is done by the **ascending nerve tracts**

- Transmission of nerve signals from the brain is done by **descending nerve tracts**.

- A cluster of nerve cell bodies located external to the central nervous system is a **ganglion** (GANG-glee-on). A ganglion can also refer to a benign cyst resembling a tumor.

- **Innervation** (in-err-VAY-shun) is the supply of nerve to a particular part of the body.

- A network of converging spinal nerve is a **plexus** (PLECK-sus). plexuses - plural. It can also refer to a network of converging lymphatic or blood vessels.

- The areas on the sensory organs (ears, taste buds, eyes, nose, and skin) that are sensitive to external stimulation are known as **receptors**. The sensory neurons transmit receptors from the stimuli to the brain where it is then interpreted.

- A **stimulus** is something that excites a nerve and results in an impulse. **Neurons** and **nerve fibers** transmit impulses- an excitatory wave.

The Reflexes

An unplanned, automated reaction to an external or internal change is called a **reflex** (REE-flecks). Reflexive actions include:

- Sneezing and Coughing

- Changes in blood pressure, breathing rate, and heart rate

- Reaction to pain.

The Neurons

The nervous system is made of **neurons** (NEW-ronz). These are the fundamental cells that facilitate communication between various body parts.

There are billions of neurons in the body. They use an electrochemical process in the transmission of nerve signals across the body. This process generates a cascade of electrical activity in the brain called **brain waves**. A variety of brain waves are generated in the course of sleep, rest, and strenuous activities.

Neuron Parts

The parts of a neuron include the cell body, one axon, numerous dendrites, and end fibers

- **Dendrites** (DEN-drytes) are processes resembling a root; they receive nerve signals and transmit them to the cell body. **Processes** are simply protrusions from the cell body.

- A process that transmits nerve signals from the neuron is referred to as **axon** (ACK-son). The length of an axon can be up to 3 feet. A significant number of neurons have a protective fatty tissue covering called **myelin sheath**.

- **Terminal fibers** are the fibers at the end of the axon which conduct nerve signal to the synapse from the axon.

- **Synapses** (SIN-apps) are gaps between a neuron and a receptor organ, or gaps between two neurons. One neuron can be made of up to a hundred synapses.

Neurotransmitters

Neurotransmitters (new-roh-trans-MIT-erz) are chemicals that facilitate the transport of signals from a neuronal synapse to the receptor organ. There are 200-

300 neurotransmitters, and they all have their unique functions. A few examples of neurotransmitters:

- **Endorphins** are natural chemicals released by the brain that helps with pain relief.

- **Acetylcholine** is produced at synapses in neuromuscular junctions and spinal cord. Its effect is produced on muscle.

- **Serotonin** is released by the brain; it influences pleasure, sleep, and hunger. A few theories also suggest that it is connected to disturbances in the mood.

- **Dopamine** is produced by the brain. Various theories suggest that it influences imbalances in thoughts and moods, and in movement disorders like Parkinson's disease.

- **Norepinephrine** is produced by the adrenal gland in response to fight or flight; it influences arousal, increases heart rate and blood pressure, and the release of glucose storage in response to stress.

Glial Cells

Glial (GLEE-ul) cells protect and support nerve cells. Their key roles include oxygen and nutrient supply to neurons, destruction and removal of damaged neurons, stabilization and anchor of neurons, and insulation of neurons from each other.

The Myelin Sheath

Glial cells make up a protective covering known as **myelin sheath** (MY-eh-lin). The white matter in the brain is formed by myelin sheath. The axons of many peripheral nerves and some portions of the spinal cord are protected by myelin sheath.

The myelinated part of nerve fibers is the white matter. Myelinated means *covered by a myelin sheath*. **Myelin** is responsible for the white color of fibers.

The gray matter is the unmyelinated part of nerve fibers. **Unmyelinated** describes the absence of myelin sheath. The gray color of the spinal cord and brain is as a result of the absence of myelin sheath.

The Central Nervous System

The components of the spinal cord include the spinal cord and brain. The vertebrae of the spinal cord, the cranial bones, provide external protection for the brain and spinal cord while internal protection is provided by the cerebrospinal fluid and the meninges

The Meninges

The spinal cord and brain are enclosed by a network of membranes called **meninges** (meh-NIN-jeez). Three layers of connective tissue make up the meninges; they include the pia mater, arachnoid membrane, and the dura mater.

The Pia Mater

This is the innermost layer of the meninges, and it is closest to the spinal cord and brain. It is composed of sensitive connective tissue that is made of an abundant supply of blood vessels. Pia means *fragile* or *tender*, mater means *mother*.

The Arachnoid Membrane

This is the middle layer of the meninges, found between the pia mater and the dura mater. This membrane looks like a spider's web. **Arachnoid** (ah-RACK-noid) means *related to spiders*.

To ensure that fluid flows freely between the layers, the arachnoid membrane is attached loosely to the remaining layers.

Cerebrospinal fluid is contained in the **subarachnoid space**, the space above the pia mater, and under the arachnoid membrane.

The Dura Mater

The external meningeal membrane is the **dura mater** (DOO-rah MAH-ter). The term dura means *hard*, and mater refers to *mother*.

- The dura mater is the lining of the interior of the skull (cranium)

- The **epidural space** is the internal part of the vertebral column. It is found between the dura mater of the meninges and the walls of the vertebral column. It is made up of supportive connective tissue and fatty tissue which protects the dura mater.

- The **subdural space** is found between the arachnoid membrane and the dura mater in the vertebral column and the cranium.

Cerebrospinal Fluid

Cerebrospinal fluid (ser-eh-broh-SPY-nal) or **spinal fluid** is secreted by specialized capillaries inside the four ventricles in the central area of the cerebrum. Cerebrospinal fluid is a colorless, transparent fluid that courses through the brain and spinal cord. It serves the following functions:

- Provision of nutrients to the brain and spinal cord by supplying nutrients and neurotransmitters to them.

- Insulation and protection of the brain and spinal cord against damage and shock.

The Parts of the Brain

The Cerebrum

The biggest and highest part of the brain is the **cerebrum** (seh-REE-brum). In addition to controlling memory, thought, emotion, and judgment, it also integrates and controls sensory and motor activities. Although the cerebrum and cerebellum seem to be the same, they are separate parts of the brain. Take note that the cerebrum is above the cerebellum. Cerebral means *related to the brain or cerebrum*. 'Cerebra'- brain, '-al'- related to.

- The outermost layer of the cerebrum is the **cerebral cortex**; it comprises of gray matter and has many deep grooves and elevated ridges.

- Within the cerebral cortex are the elevated ridges of gray matter called the **gyri**. Gyrus - singular.

- The fissures in the cerebral cortex are called **sulci**. In this context, a **fissure** is a normal deep furrow. In the skin, fissures could refer to sores that look like cracks.

The Cerebral Hemispheres

Two cerebral hemispheres are created from the division of the cerebrum, and the **corpus callosum** connects them at the center.

- Most functions on the right side of the body are controlled by the **left cerebral hemisphere**. This is why the motor and sensory becomes imbalanced if there is damage to the left hemisphere.

- Most functions on the left side of the body are controlled by the **right cerebral hemisphere**. This is why the motor and sensory becomes imbalanced if there is damage to the right hemisphere.

- This complex system is as a result of the **intersection of nerve fibers** in the brain stem.

The Cerebral Lobes

The subdivision of each cerebral hemisphere produces cerebral lobes, which are in pairs. The lobes are named after each enclosing cranial bone.

- The **frontal lobe** is responsible for the control of memory, function, and behavior.

- The **occipital lobe** is responsible for controlling vision.

- Impulses from sensory receptors in the skin, muscles, and tongue are received and interpreted by the **parietal lobe.**

- Creation, Storage, and recall of new information are controlled by the **temporal lobe**. It also controls the senses of smell and hearing.

The Thalamus

The transmission of nerve signals from and to the cerebrum and sense organs result in the generation of sensations, and this is done by the **thalamus** (THAL-ah-mus), a structure found under the cerebrum.

The Hypothalamus

This structure is found under the thalamus. The **hypothalamus** (high-poh-THAL-ah-mus) performs key regulatory functions in the body.

The Cerebellum

- This is the second largest part of the brain; the **cerebellum** (ser-eh-BELL-um) is found at the back of the head and posterior to the posterior part of the cerebrum.

- Incoming impulses concerning joint movements, body positions, and muscle tone are received by the cerebellum. The information is then transmitted to the various parts of the brain responsible for the control of muscle movement.

- The basic roles of the cerebellum include the maintenance of normal body posture, coordination of movement and maintenance of balance.

The Brainstem

This is the stalk-like part of the brain that forms a connection between the spinal cord and the cerebral hemisphere. It is divided into three: medulla oblongata, pons, and midbrain.

- The **medulla oblongata** (meh-DULL-ah ob-long-GAH-tah) is found at the bottom of the brain stem and it is linked to the spinal cord. It controls the reflexes for vomitting, coughing, swallowing, and sneezing. It is also responsible for basic survival functions that include respiratory muscles, blood pressure, and heart rate.

- The **pons** (PONZ) and **midbrain** provide a pathway for the transmission of impulses from the lower and higher brain centers. In addition, they control reflexive eye and head movement in reaction to auditory and visual stimuli. The Latin word, pons means *bridge*.

The Spinal Cord

- This long and delicate tubular structure starts at the end of the brain stem and extends downward, almost to the end of the spinal cord.

- All the nerves that control the limbs and lower body parts are contained in the spinal cord. It provides a pathway for the transmission of impulses to and from the brain.

- Meninges and cerebrospinal fluid protect and surround the spinal cord.

The Peripheral Nervous System

The components of the peripheral nervous system are 31 pairs of peripheral nerves branching outwards from the spinal cord and 12 pairs of cranial nerve branching outwards from the brain. The term **peripheral** refers to body parts that are located at a further distance from the center of the body.

Transmission of nerve impulses to and from the central nervous system is done by the 3 kinds of peripheral nerves; they include: somatic, sensory, and autonomic nerve fibers.

- Otherwise called **motor nerve fibers**, the **somatic nerve fibers** transmit impulses that direct voluntary movement of muscles.

- Stimuli from outside the body, like the taste of something, are sent to the **sensory neurons** which conveys the impulse to the brain where they are interpreted.

- Messages are transported from the autonomic nervous system to the organs and glands by the **autonomic nerve fibers**.

The Cranial Nerves

There are 12 pairs of cranial nerves that emerge from the base of the brain. One pair consists of two nerves that are similar in structure and function, and one nerve from a pair innervates one side of the body. They are classified according to roman numerals and are named according to the functions performed or the area they are located.

The Peripheral Spinal Nerves

- There are 31 pairs of peripheral spinal nerves

which are collectively named according to the part of the body that they serve.

- According to each part, they are grouped numerically. C1 through C8 for the cervical nerves, TI through T12 represents the thoracic nerves, L1 through L5 for lumbar nerves, and S1 through S5 for sacral nerves.

- In a few instances, spinal nerve forms a network of **plexus** to innervate a particular area of the body. The first 4 lumbar nerves form the **lumbar plexus** which innervates the lower back.

The Autonomic Nervous System

There are two broad divisions of the autonomic nervous system. The first division is made up of the **sympathetic nerves**, and the other division is made up of the **parasympathetic nerves**.

Involuntary activities like heart rate and blood pressure are controlled by the **autonomic nervous system**. The actions of one division complement the actions of the other division and this helps to ensure homeostasis. The process of maintaining a stable internal body environment is what is known as **homeostasis**.

The parasympathetic nerves ensure the normal functioning of the body following a physiological response to stress. In addition, they control regular functioning of the body in situations that do not require physical or mental effort.

On the other hand, the body is prepared for stress and emergencies by the **sympathetic nervous system**; this system increases the heart rate, flow of blood to muscles, and breathing rate. The stimulation of these neurons sets off the flight or fight response, the body's programmed response to a dangerous situation, whether imaginary or real.

Chapter 7: The Cardiovascular system

The word cardiovascular means pertaining to the heart and blood vessels. Thus, the cardiovascular system is made up of the heart, the blood, and the vessels through which the blood is circulated around the body.

Cardi/o: heart, -vascul: blood vessels, and -ar: pertaining to.

Blood

Blood is a fluid tissue that consists of red blood cells (that transports oxygen), white blood cells (that help the body fight infections), and platelets (that helps stop blood loss during an injury). All these cells are suspended in a liquid medium known as the plasma.

The blood also helps to remove waste products from the body by transporting dissolved metabolic waste products to the kidneys and to remove carbon dioxide from the lungs. In addition, hormones, nutrients, and other materials are distributed to various parts of the body by the blood.

Structures of the Cardiovascular System

The main components of the cardiovascular system are:

- The heart

- The blood vessels, and

- The blood.

The Heart

The heart is a hollow muscular organ. It is is divided into four chambers, and located within the thoracic cavity, just above the stomach and between the lungs. The pumping force of the heart is efficient enough to ensure the flow of blood all through the body.

The Pericardium

The double-walled membrane that surrounds the heart is known as the **pericardium** or the **pericardial sac**.

Peri: surrounding, -cardi: heart, and -um is a singular noun ending.

A **membrane** is a thin layer of pliable tissue that serves as a covering or envelope to a body part.

The pericardium consists of:

- The **parietal pericardium**, a fibrous sac found around the heart that serves as a protective cover for it.

- The **visceral pericardium**, the innermost layer of the pericardium that has direct contact with the heart (also known as the **epicardium**).

- The **pericardial fluid**, a lubricant trapped between the two layers of the pericardium that prevents friction caused by the heartbeats.

The Walls of the Heart

There are three layers that make up the walls of the heart. They are:

- The **epicardium** (ep-ih-KAR-dee-um): This is the outermost layer of the heart (and the inner layer of the pericar-dium).

  Epi: upon, -cardi: heart, and -um is a singular noun ending.

- The **myocardium** (my-oh-KAR-dee-um): This is the middle layer and the thickest of the heart's three layers. It consists of specialized muscles capable of contraction and relaxation, thereby causing the pumping action of the heart.

My/o: muscle, -Cardi: heart, and -Um is a singular noun ending.

- The **endocardium** (en-doh-KAR-dee-um): The endocardium is the innermost layer of the heart and it is made up of squamous epithelial tissue, which serves as the inner lining of the heart. The surface of the endocardium has direct contact with the blood as it flows in and is pumped out of the heart.

  Endo: within, -cardi: heart, and -um is a singular noun ending.

Blood Supply to the Myocardium

The power engine of the heart known as the **myocardium**. It needs a steady supply of nutrients and oxygen. The **coronary arteries** (KOR-uh-nerr-ee) perform the role of supplying these. Any damage to the coronary artery that supplies a portion of the mycardium will cause the death of that affected region. The veins on the other hand help to remove waste products from the myocardium.

The Chambers of the Heart

The heart is made up of four chambers, each performing a specialized function:

- **The atria** (AY-tree-ah): The upper chamber of the heart is known as the atria, It is separated by the **Interatrial septum** into right and left atrium. Blood coming from the body enters the atria, thus it is the receiving chambers of the heart.

- **The ventricles** (VEN-trih-kuhls): The two lower chambers of the heart, separated from the atria by **valves** and divided by the **interventricular septum**. The ventricular walls are thicker than those of the atria due to the pumping action of the ventricles.

The Valves of the Heart

There are four valves that determine the blood flow through the heart. They are:

- **The tricuspid valve** (try-KUS-pid): This valve controls the portal between the right atrium and the right ventricle.

- **The pulmonary semilunar valve** (PULL-mah-nair-ee sem-ee-LOO-nar): Controls the flow of blood between the pulmonary artery and the right ventricle.

- **The mitral valve** (MY-tral): Controls the opening between the left atrium and left ventricle.

- **The aortic semilunar valve** (ay-OR-tick sem-ee-LOO-nar): This is found between the aorta and the left ventricle.

Pulmonary and Systemic Circulation

The human heart has a double circulatory system called the pulmonary and the systemic circulation.

Pulmonary circulation:

Pulmonary circulation occurs between the heart and the lungs alone.

- Deoxygenated blood is transported out of the right ventricle into the lungs through the pulmonary arteries. (Only the pulmonary artery carries deoxygenated blood).

- Deoxygenated blood deposits carbon dioxide from the body into the lungs in exchange for inhaled oxygen.

- On the other hand, the pulmonary veins transport oxygenated blood from the lungs into the left atrium of the heart.

Systemic circulation:

Systemic circulation involves the flow of blood from the heart to all parts of the body except the lungs.

- Oxygenated blood is pumped to all parts of the body through the left ventricle of the left atrium.

- The veins of the body carry deoxygenated blood from the body through the right atrium.

The Heartbeat

The heartbeat is observed due to the pumping action of the heart. Blood is pumped by the contraction and relaxation of the cardiac muscles.

The rate of heartbeat is determined by **electrical impulses** conducted by nerves which stimulate the myocardium of the heart. The **sinoatrial** (SA) node, **atrioventricular** (AV) node, and the **bundle of His** initiates the impulse that stimulates the myocardium.

The Sinoatrial (SA) Node

The SA node is located close to the entrance of the superior vena cava in the posterior wall of the right atrium. It is often known as the natural pacemaker because it initiates and determines the rhythm and rate of the heartbeat.

The Atrioventricular (AV) Node

The impulse from the SA node is conducted to the AV node. The AV node is located at the base of the right atrium, close to the interatrial septum. The AV node helps to transmit electrical impulses further towards the bundle of His.

The Bundle of His

This is a group or bundle of muscle fibers found within the interventricular septum. The bundle of His conducts electrical impulses to the right and left ventricles and Purkinje fibers.

Purkinje fibers are a group of specialized conductive fibers found within the walls of the ventricles. They help to conduct electrical impulses to the cells of the ventricles.

Electrical Waves

The electrical activities of the heart can be visualized in the form of wave (PQRST) movements using an electrocardiogram or monitor.

- The P wave is as a result of the contraction of the atria.

- The QRS complex stimulation indicates contraction of the ventricles

- The T wave represents the relaxation or recovery of the ventricles.

The Blood Vessels

The blood vessels comprise of the: arteries, capillaries, and veins.

Arteries

Arteries are large blood vessels that carry oxygenated blood from the heart to other parts of the body. The arterial blood is well oxygenated; hence, it is bright red in color.

The largest artery is **aorta** (ay-OR-tah) that originates from the left ventricle of the heart to form the trunk of the arterial system. Blood is carried from the heart towards the head through the **carotid** (kah-ROT-id) arteries, thereby supplying the brain with oxygen-rich blood. Shortage of blood supply to the brain could lead to stroke.

Arterioles (ar-TEE-ree-ohlz) are the smaller arteries that help supply blood to the capillaries. The rate of blood flow at the capillary bed is greatly reduced.

Capillaries

Capillaries (KAP-uh-ler-eez) are the smallest blood vessels in the body, and they are highly permeable. The capillaries branch out to form a network vessel that can deliver oxygen and dissolved nutrients directly to cells of vascularized tissues.

Veins

The vein carries deoxygenated blood from the tissues to the heart. **Venules** (VEN-youls), just like the arterioles, are the smallest veins and they combine to form the larger veins. They collect blood from cells of tissues and carry them to the heart.

Superficial veins are found close to the body surface while **deep veins** are located within the tissues.

The Venae Cavae

The venae cavae (VEE-nee KAY-vee) are veins that collect blood from all the other veins and return it to the heart. The superior vena cava transports blood from the upper portion of the body while the inferior vena cava transports blood from the lower portion of the body. They are the two largest veins of the body.

Pulse

Contraction of the heart causes a rhythmic pressure against the walls of the artery. Such pressure is known as a **pulse**.

Blood Pressure

The amount of systolic and diastolic pressure exerted by the heart is known as the **blood pressure**.

- **Systolic pressure** (sis-TOL-ick): This is the highest pressure against the walls of an artery. It occurs when the walls of the ventricles contract, thereby pumping blood throughout the body.

- **Diastolic pressure** (dye-ah-STOL-ick): This is the relaxation phase of the ventricle, and lowest pressure against the walls of an artery.

Blood

Blood is a tissue in fluid form, and it is composed of 45% formed elements (the cells) and 55% plasma.

Plasma

All the cells of the body are suspended in a straw-colored fluid known as the plasma (PLAZ-mah). The plasma also contains dissolved minerals, proteins, nutrients, hormones, and waste products. About 91% of the plasma is made up of water while the other 9% consists of clotting factors and proteins.

Serum (SEER-um) is just like plasma except that clotting factor is absent.

Cells of the Blood

The blood consists of three cells which are:

- The erythrocytes,

- The leukocytes, and

- The thrombocytes.

Erythrocytes (Red blood cells or RBC)

The red blood cells, also known as the **erythrocytes** (eh-RITH-roh-sights), are produced in the red bone marrow. The RBC contains hemoglobin (hee-moh-GLOH-bin) which helps the RBC to transport oxygen to the tissues of the body.

Leukocytes (White blood cells or WBC)

The white blood cells or Leukocytes (LOO-koh-sites) helps the body to fight infections. There are five major types of leukocytes:

- **Neutrophils** (NEW-troh-fills) fight infections caused by bacteria and other microorganisms by phagocytosis.

- **Basophils** (BAY-soh-fills) cause allergic symptoms by releasing its granules (degranulation) during an infection.

105

- **Eosinophils** (ee-oh-SIN-oh-fills) primarily help to fight parasitic infections such as parasitic worms. They also contain coarse granules which when released can cause allergic symptoms as well.

- **Lymphocytes** (LIM-foh-sights) are produced in the bone marrows, but their maturity takes place in the lymph nodes and spleen. They act primarily against viral infections.

- **Monocytes** (MON-oh-sights) also fight infections by phagocytosis, and they can act as antigen-presenting cells.

Thrombocytes (Platelets)

Thrombocytes (THROM-boh-sights) are the smallest cells of the blood. Their primary role is the formation of blood clots during a vascular injury in order to stop blood loss.

Blood Types

The red blood cells are grouped based on the presence or absence of specific antigens on the surface of the RBC.

An **antigen** is any substance that is foreign to the body. Based on this premise, the blood is grouped into four main blood groups:

- A,

- AB,

- B, and

- O

The Rhesus factor (Rh factor)

The **Rhesus factor** of an individual is determined by the presence or absence of the Rh antigen on the surface of the RBC.

The Rh cross-matching test is a key test before any blood transfusion can be carried out.

Blood Gases

There are three main blood gases that are normally dissolved in the blood. They are:

- Oxygen (O_2),

- Carbon dioxide (CO_2), and

- Nitrogen (N_2).

Chapter 8. Endocrine system

Production of hormones that collaboratively maintain homeostasis is the main function of the endocrine system. A stable internal environment is maintained by the blood through the process of **homeostasis** (hoh-mee-oh-STAY-sis). The term homeo means *constant* while stasis means *control*.

Chemical messengers that are secreted directly into the blood are called **hormones**. Since they are secreted directly into the blood, they are able to spread across organs and cells across the body.

Each hormone plays its own unique role in the regulation of organs and cells, and their levels are measured via urine or blood tests.

Structures of the Endocrine System

Endocrine (EN-doh-krin) glands produce hormones. They are ductless glands. The term 'endo' means *internal*, and '-crine' means *production*.

The endocrine system is made up of 13 important glands

- One thyroid gland

- One pancreas

- One pineal gland

- One thymus

- One pituitary gland (subdivided into anterior and posterior lobes)

- Two gonads (may be a pair of testes in men, or a pair of ovaries in women)

- Two adrenal glands

- Four parathyroid glands

The Thyroid Gland

This gland is shaped like a butterfly and is located under the thyroid cartilage, on both sides of the larynx. A major role of the thyroid gland is the regulation of the body metabolism. The collective name for all the processes by which the body utilizes nutrients and the rate of utilization is **metabolism**. The thyroid gland secretes hormones which also affect the functioning and development of the nervous system.

The metabolic rate, development and functioning of many systems in the body is controlled by the teo main thyroid hormones which are:

- triiodothyronine (T3) (try-eye-oh-doh-THIGH-roh-neen)

- thyroxine (T4) (thigh-ROCK-seen)

The **Thyroid Stimulating Hormone** (TSH) which is produced in the anterior lobe of the pituitary gland regulates the rate at which theses secretions are produced.

Another hormone called **calcitonin** (kal-sih-TOH-nin) is secreted by the thyroid gland and it works together with the parathyroid hormone to reduce the levels of calcium in tissues and blood by directing the storage of calcium to the teeth and bones.

The Pancreas (Pancreatic Islets)

As a feather-shaped organ that lies posterior to the stomach, the **pancreas** (PAN-kree-as) serves as a part of the endocrine and digestive systems.

- The endocrine functions of the pancreas are carried out by the **pancreatic islets** (pan-kree-AT-ick EYE-lets). An **islet** or **island** is a small distinct collection of a particular kind of tissue within a bigger mass of tissue type.

- These islets carry out the endocrine function of regulating glucose metabolism and levels of blood sugar across the body. **Blood sugar** or **glucose** (GLOO-kohs) is the simplest form of energy that is utilized in the body.

- The islets also secrets other hormones such as **insulin** (IN-suh-lin) which is produced by the beta cells of the pancreas when there is excess glucose in the blood. There are 2 ways by which insulin acts

  o If there is a depletion of energy, insulin facilitates the entry of glucose into cells so that they can be utilized as a source of energy

  o If there is an increased level of glucose, insulin activates the liver to store glycogen which is gotten from the modification of glucose.

- **Glucagon** (GLOO-kah-gon) is produced by the alpha cells of the pancreas when there are reduced levels of glucose in the blood. Glucagon raises the level of blood sugar by stimulating the liver to convert stored glycogen act to glucose which is then released into the blood.

The Pineal Gland

- The **pineal (PIN-ee-al) body** otherwise called the **pineal gland** is a tiny endocrine gland that is found in the central part of the brain.

- The pineal gland produces hormone melatonin (mel-ah-TOH-nin) that regulate the cycle of sleep and wakefulness in the circadian rhythm. The physiological activities that happen in the course of 24 hours are known as **circadian rhythm**.

The Thymus

- The **thymus** (THIGH-mus) is found close to the central portion of the anterior part of the thoracic cavity. It lies above (superior to) the heart and behind (posterior to) the sternum.

- It serves its role as an endocrine organ by producing a hormone that plays a key role in the immune system. The hormone secreted by the thyroid is called **thymosin**.

- **Thymosin** (THIGH-moh-sin) is an important hormone involved in the regulation of immunity. It helps immature lymphocytes to develop into mature T cells.

The Pituitary Gland

Otherwise called **hyphophysis**, the **pituitary (pih-TOO-ih-tair-ee) gland** is a pea-sized gland that has 2 divisions, the anterior lobe and the posterior lobe. Both lobes dangle from a stalk-shaped structure found under the hypothalamus (in the brain).

The hypothalamus is an important organ in the nervous system that secretes hormones involved in the regulation of various functions in the body.

Functions of the Pituitary Gland

The functions of the other endocrine glands are regulated by hormones secreted by the pituitary gland; and the secretion of theses regulatory hormones is the main function of the pituitary gland.

The hypothalamus secretes neurohormones which stimulates the actions of the pituitary gland. This system creates a regulatory mechanism that ensures an adequate quantity of each hormone in the bloodstream.

Secretions of the Pituitary Gland: Anterior Lobe

The hormones secreted by the anterior lobe of the pituitary gland include:

- The **somatotropic hormone**, simply referred to as the **growth hormone** (GH). This hormone controls the development of muscle, bone, and other tissues in the body. 'soma'- body, '-tropic'- attracted to.

- The **adrenocorticotropic (ah-DREE-noh-kor-tih-coe-TROP-ik) hormone** (ACTH) regulates the secretion and development of the adrenal cortex. 'adreno'- adrenal, 'cortico'- cortex, 'trop'- change, '-ic'- related to.

- The **interstitial (in-ter-STISH-al) cell-stimulating hormone** (ICSH) activates ovulation in women. It activates the production of testosterone in men.

- The **follicle-stimulating hormone** (FSH) activates the development and secretion of eggs (ova) in a woman's ovaries. It activates the manufacture of sperm in the testes.

- The **melanocyte-stimulating (mel-LAN-oh-sight) hormone** (MSH) causes an elevation in the melanin produced by the melanocytes, which is responsible for the dark color of the skin. During pregnancy, there is an increase in the production of MSH.

- The **luteinizing (LOO-tee-in-eye-zing) hormone** (LH) activates ovulation in women. In men, it activates the production of testosterone.

- **Prolactin** or **lactogenic hormone** (LTH) is responsible for the activation and maintenance of breast milk secretion in the mother, after delivery. 'lacto'- milk, 'gen'- production, '-ic'- related to.

The secretion of all these hormones by the thyroid gland is regulated by the **thyroid-stimulating hormone** (TSH)

Secretions of the Pituitary Gland: Posterior Lobe

- **Oxytocin** (ock-see-TOH-sin) (OXT) activates the contraction of the uterus during delivery. 'oxy'- swift, '-tocin'- labor. This hormone activates milk flow from the mammary gland and postpartum hemorrhage, after delivery. A synthetic variant of oxytocin that induces or increases labor is called **pitocin**.

- **Antidiuretic (an-tih-dye-you-RET-ick) hormone** (ADH) is produced by the hypothalamus and stored in the pituitary gland. It regulates blood pressure by decreasing the quantity of water excreted by the kidneys. On the other hand, a diuretric is given to raise the quantity of secreted urine.

The Gonads

Gonads (GOH-nadz) are glands which produce gamete. The gonads are the testes in males and the ovaries in females.

Functions of the Gonads

The gonads produce the hormones that handle the growth and maintenance of secondary sexual characteristics which develop in the course of puberty. The features that differentiate males from females but have no direct connection to reproduction are known as **secondary sex characteristics**.

Puberty (PYU-ber-tee) is the series of events that result in physical changes of a child's body into an adult body that can effectively reproduce. It is characterized by the development of secondary sex characteristics, maturation of male and female genitals, and by the onset of menstruation in girls. The average age for puberty is 11 for boys, and 12 for girls.

Precocious puberty is the premature occurrence of the physical changes of puberty; in boys, it happens at age 9 and in girls, age 8. The term **precocious** means extremely early in onset or development.

Secretions of the Gonads

A reproductive cell is called a **gamete** (GAM-eet). It is the **egg** (ovum) in girls and the **sperm** in boys.

- **Estrogen** (ES-troh-jen) is produced by the ovaries and is significant to the maturity and maintenance of the secondary sex characteristics in females. It also regulates the menstrual cycle.

- **Progesterone** (proh-JES-ter-ohn) is released in the second phase of the menstrual cycle in the ovary by the **corpus luteum**. It plays a key role in preparing the uterus for pregnancy.

- **Testosterone** (tes-TOS-teh-rohn) is secreted by the testes and the adrenal cortex. It is a steroid hormone that stimulates the development of secondary sex characteristics in males.

- **Androgens** (AN-droh-jenz) are similar to testosterone and are sex hormones which are produced by the adrenal cortex, gonads, and fat cells. Although androgens are found in males and females, they stimulate the growth and maintenance of secondary sex characteristics in men.

- **Gonadotropin** (gon-ah-doh-TROH-pin) is the hormone that activates the gonads. 'gonado'- gonad, '-tropin'- stimulate.

The Adrenal Glands

These glands are found on top of each kidney and that is why the adrenal glands are also called the **suprarenal glands**. The adrenal gland is divided into two parts, the outermost part, or the adrenal cortex; and the inner part, or the adrenal medulla. These parts have their own unique function.

Regulation of the levels of electrolyte in the body is one of the key roles of the adrenal glands. **Electrolytes** (ee-LECK-troh-lytes) are minerals such as potassium and sodium that are present in the blood.

The adrenal gland is also important in the regulation of metabolism and control of the response of the sympathetic nervous system to stress.

Secretions of the Adrenal Cortex

The hormones produced by the adrenal cortex include:

- **Corticosteroids** (kor-tih-koh-STEHR-oidz) are secreted by the adrenal cortex. It is a collective name for all the steroid hormones. The synthetic variant is produced and administered as drugs.

- **Cortisol** (KOR-tih-sol) is otherwise referred to as **hydrocortisone** and it is a corticosteroid that not only controls the metabolism of fats, protein, and carbohydrates in the body, but also has anti-inflammatory properties.

- **Aldosterone** (al-DOSS-ter-ohn) is a corticosteroid responsible for the control of water and electrolyte levels in the body. It does this by increasing the excretion of potassium and reabsorption of sodium by the kidneys. The process of reabsorption involves the return of substances into the blood.

- **Androgens** are the sex hormones produced by the adrenal cortex, and other cells like fat cells, and the gonads.

Secretions of the Adrenal Medulla

The adrenal medulla secretes two hormones:

- **Norepinephrine** (nor-ep-ih-NEF-rin), which is released by the adrenal medulla as a hormone and by the sympathetic nervous system as a neurohormone. It plays a significant function in the fight or flight response in the body which happens through the stimulation of muscle contractions, elevation of blood pressure, and increasing the force of the heartbeat.

- **Epinephrine** (ep-ih-NEF-rin) is otherwise called **adrenaline**. It is responsible for stimulating the response of the sympathetic nervous system to emotional stress like fear, and physical stress as well. It increases the heart beat and elevates blood pressure. Additionally, it inhibits the release of insulin and stimulates the release of glucose by the liver

The Parathyroid Glands

There are four parathyroid glands and each one of them is as big as a grain of rice. They are buried within the posterior part of the thyroid gland.

The key role of the parathyroid gland is in the control of the levels of calcium in the body. The level of calcium in the body influences the activities of the nervous and muscular systems. The hormones produces by the parathyroid gland is called the parathyroid hormone.

The **parathyroid hormone** (PTH) which functions together with **calcitonin**, the hormone produced by the thyroid gland. The regulation of calcium levels in tissues and blood is a joint effort of calcitonin and the parathyroid hormone.

If the level of parathyroid hormone in the body is elevated above the normal levels, it can result in the release of stored calcium within the teeth and bones, thus increasing the levels of calcium in the blood.

Specialized Types of Hormones

There are several hormones that do not conform to the typical definition of hormones, probably because they are not produced directly into the blood by endocrine glands, or due to their chemical structure, these hormones are **specialized hormones**.

Steroids

Steroids (STEHR-oidz) are an expansive family of fat-soluble substances that are chemically related to hormones. Testosterone, anti-inflammatory drugs, and cholesterol are examples of steroids.

The endocrine gland produces steroids, and they can also be manufactured in form of medications that alleviate inflammation and swelling in medical conditions like asthma.

Anabolic steroids (an-ah-BOL-ick STEHR-oidz) are artificially produced substances that share the same chemical structure with the sex hormones of males. They are important in the treatment of hormonal imbalances in males, and they facilitate the replacement of muscle mass that may have been reduced in the course of a disease. These steroids are illegally used by sportsmen to bulk up their muscles; this is a harmful practice that can permanently damage the organs in the body.

Hormones Secreted by Fat Cells

It is uncommon to regard adipose tissue as endocrine glands but various studies have discovered that fat cells actually produce a minimum of one, and maybe even more hormones that are involved in maintaining the health and balance of the body. An example of the hormone secreted by adipose tissue is **Leptin** (LEP-tin), a protein hormone that is important in the control of appetite.

Adipose tissue releases leptin which is transported by the bloodstream to the brain, it stimulates the hypothalamus which represses hunger, and uses up the fat stored in fat cells.

Neurohormones

Neurohormones (new-roh-HOR-mohnz) are manufactured and released by nerve cells in the brain, and not the endocrine glands. It is then transported by the bloodstream to tissues and organs. The hypothalamus secretes neurohormones which regulate the secretion of hormones produced by the pituitary gland.

Chapter 9: The Integumentary System

The skin and all associated tissues make up what is called the 'Integumentary system' (in-teg-you-MEN-tah-ree) and they carry out essential roles in protecting the body. "Integument" means *'to encompass'*

- The skin encloses every organ in the body as the most exterior defensive structure.

- It preserves the body's moisture.

- Undamaged skin shields the body from pathogens.

- It receives the majority of 'touch' sensations.

- It absorbs Vitamin D from sunlight and blocks its unsafe UV radiation.

- It is the largest organ of the body.

The integumentary system has associated components; sweat glands, sebaceous glands, nails, and hair.

- Hair restricts heat depletion from the body

- Nails protects the bones at the tip of the toes and fingers

- The sebaceous (seh-BAY-shus) glands produce an oil called **sebum** that inhibits the multiplication of microbes on the skin because of its slight acidity.

- The sweat glands modify body temperature and fluid load by producing sweat containing bodily waste.

The Structures of The Skin

The Skin consists of 3 layers:

- The Epidermis (ep-ih-DER-mis)

- The Dermis (DER-mis)

- The Subcutaneous (sub-kyou-TAY-nee-us) layers

'Cutaneous' means *'related to the skin'* and 'sub' means *'under'*

'Epi' means *'above'* and 'derm' means *'skin'*

The Epidermis

This is the most exterior stratum with lots of epithelial

tissues (that shields every surface of the body). It has no connective tissues or blood vessels, making it reliant on layers below for nutrition. Other complex tissues include;

- **Squamous epithelial tissue** (SKWAY-mus) meaning *'scale-like'* tissue is the topmost layer of the epidermis and it is made up of rough, leveled cells that are regularly shed.

- **Basal layer** (BAY-suhl) is the lowest layer of the epidermis that makes new skin cells and moves them above. When these new cells die, they change into keratin-filled cells.

- **Keratin** (KER-ah-tin) is a fibrous protein that repulses water. It is of two kinds; soft (primary constituent of the epidermis) and hard (component of hair and nails).

- **Melanocytes** (MEL-ah-noh-sights) are specialized cells in the basal layer of the epidermis. They help in melanin production.

- **Melanin** (MEL-ah-nin) is a dark pigment. It establishes what color the skin and hair will be. It plays a vital role in shielding the skin from unsafe UV radiation from the sun. It is also responsible for unique spots like age spots, moles and freckles.

The Dermis

This is the thick layer of living cells right below the epidermis. It has blood and lymph vessels, connective tissues, and nerve endings (receptors for touch, pain, pressure, and temperature). Sebaceous glands, sweat glands, nails, and hair follicles are also present in the dermis. It is also called **the corium**.

Tissues within the Dermis

- **Collagen** (KOL-ah-jen) is a protein that means **adhesive,** and it is found in cartilage, tendons, ligaments, and bones. It has great tensile strength.

- **Mast cells** secrete substances; histamine and heparin that react to allergies, wounds, and infections.

- **Heparin** (HEP-ah-rin) inhibits blood clotting

- **Histamine** (HISS-tah-meen) reacts to allergy-causing substances and causes all signs of inflammation.

The Subcutaneous Layer

This layer is below every other layer of skin and consists of loose connective tissue and **adipose** (AD-ih-pohs) or **fatty tissue**. It attaches the skin to the directly underlying muscles.

- **Cellulite** (different from the regular word cellulitis) is a regular word used to refer to dimpled fat gathered around the bum and thighs

- **Lipocytes** (LIP-oh-sights) are also called **fat cells**. They are mostly situated in this layer and they make and store lots of fat.

The Sebaceous (seh-BAY-shus) Glands

- Located in the dermis layer of the skin.

- The sebum (SEE-bum) they produce is secreted via channels that link the hair follicles through which it goes on to lubricate the skin surface.

- The breasts are an adaptation of sebaceous glands.

The Sweat Glands

They are minute, curled up glands that secrete sweat and are on virtually every body surface. They are even more abundant on our palms, soles, armpits, and foreheads. They are also called **sudoriferous glands**.

- The skin exterior has pores through which sweat is secreted to the surface.

- Sweat is 99% water and 1% salt and bodily waste.

- **Perspiration** is another word for sweat and through it, the body expends surplus water.

- Sweating also lowers the body's temperature. Sweat in contact with microbes on the skin's exterior is what causes body odor.

- The synthesis and secretion of sweat is called **Hidrosis** (high-DROH-sis).

The Hair

Hair is made up of hair fibers. These are cylindrical structures comprised of dead, compact cells full of hard keratin.

- **Hair follicles** (FOL-lick-kulz) grip the fiber base. Straight, coily or curly hair is dependent on the shape of follicles.

- Despite it being dead, the cells at the root multiply at a fast rate and nudge older cells above during which these old cells become hard and pigmented.

- The muscle fibers responsible for goosebumps (the raising of some areas of the skin as a reaction to cold or fear), and erect hair are the

arrector pili (ah-RECK-tor PYE-lye)

The Nails

The fingernail or toenail is also called **unguis** (UNG-gwis). Every nail is made up of;

- The clear part, **NAIL BODY**, that is meshed to the tissues below. Hard, epidermal cells filled with keratin make up the nail body.

- The nail bed containing blood vessels responsible for the usual pink color of the nail. It attaches the nail body to the connective tissues below.

- The **lunula** (LOO-new-lah), meaning **little moon**, is a semi-crescent like area of the nail root. It is most prominent in the nail of the thumb. The lunula is the part of the nail where keratin cells are continually made.

- The **cuticle** is a slim strip of epidermis which joins the nail exterior before the root of the nail. It safeguards the newly made keratin cells.

- The free edge is the part of the nail that protrudes just past the tip of the toe or finger. It is not joined to the nail bed.

- The nail root holds fast the nail to the toe or finger by being rooted to a notch in the skin.

Chapter 10: The Lymphatic and Immune Systems

The lymphatic system has 3 primary roles:

- Utilizing the lacteals in the small intestines to absorb fat and vitamins (soluble in fat) that otherwise cannot be conveyed via the blood vessels.

- Excreting bodily waste and aiding the immune system to rid the body of microbes.

- Taking purified lymph back to the veins at neck base.

Absorption of Fats and Fat-Soluble Vitamins

Lacteals along with blood vessels comprise of little bulges called **villi** which aid in the digestion of food in the small intestine.

- The cells of the lacteals (LACK-tee-ahlz) alter fats then take them back to the veins to supply the body with nourishment.

- The fat, fat-soluble vitamins, and other nutrients are taken up by the blood vessels from absorbed food for nourishment.

Interstitial Fluid and Lymph Creation

Interstitial fluid (in-ter-STISH-al) also called **tissue fluid** is plasma from blood that seeps through the spaces in between cells

- It conveys oxygen, hormones, and nutrients throughout the body

- It carries waste and protein from cells, and most (90%) of the fluid goes back to the bloodstream.

Lymph (limf) is clear and water-like consisting of proteins and electrolytes. It makes up what is left (10%) of the tissue fluid. It is crucial to the lymphatic system especially in its role in the immune system.

- Lymph removes protein, waste, microbes and abnormal (cancerous) cells from remnants in the intercellular spaces.

- It flows through tiny capillaries to bigger vessels (found much deeper in tissues) in its direct movement up to the neck.

- At this point, the lymph then takes on even more involved functions in the immune system.

Structures of the Lymphatic System

These structures work together even in various other systems.

Lymphatic Circulation

The lymphatic system is usually called a secondary circulatory system because of how identical it is to the blood circulatory system. However, there are quite a few disparities that should be taken note of;

- Unlike the blood circulatory system which flows in a loop from the heart and back, lymph follows a direct, one-way trip upwards up till the base of the neck.

- The pump organ of the blood circulatory system is the heart while there is none for the lymphatic system, hence it has to rely on muscles to propel the fluid up.

- The kidneys screen and purify the blood, and the Genito-urinary system removes metabolic waste. Whereas, lymph nodes found among lymphatic vessels screen and purify lymph.

- Blood vessels are easily seen because of the red color of blood as opposed to lymphatic vessels which are harder to see because of the clear, colorless nature of lymph.

Lymphatic Capillaries

These are minuscule tubes situated just close to the body's exterior. The walls of these tubes are very thin (one cell thick) and the cells get disjointed for a short while for lymph to flow into the capillary. The closure of the capillaries create the force that propels the fluid upwards.

Lymphatic Vessels and Ducts

Similar to veins, lymphatic vessels have valves to hinder lymph from reverse flowing. The big lymphatic vessels ultimately come together to make two ducts, both of which drain a particular area of the body into the veins.

- The **right lymphatic duct** drains the right side of the: upper part of the body, head, and neck, and the right arm, into the right subclavian vein (a section of the major vein in the arm).

- The **thoracic duct** (the biggest lymphatic vessel) drains the left side of the: upper part of the trunk, head and neck, left arm, the lower part of the torso and the two legs into the left subclavian vein.

Lymph Nodes

These are little pod-like structures that have **microcidal lymphocytes** (microbe-killing). Unclean lymph enters the nodes where microbes and abnormal cells are killed and purification occurs. Thereafter, the strained lymph flows out of the node to resume its trip to the veins.

There are about 400-700 nodes found amongst the lymphatic vessels, and about half of them can be found in the abdomen and the others scattered across the body on off-shoots of bigger lymphatic vessels. The following are 3 main classes of lymph nodes that are an exclusion and are instead termed after the locations they are found in;

- **Cervical (SER-vih-kal) lymph nodes** found at the sides of the neck

- **Axillary (AK-sih-lar-we) lymph nodes** found beneath the arms (armpits)

- **Inguinal (ING-gwih-nal) lymph nodes** found in the groin and lower abdomen.

Lymphocytes

Lymphocytes (LIM-foh-sights), also called **lymphoid cells**, are an important component of the immune system produced by the bone marrow. The 3 kinds are; T cells, B cells, and Natural killer cells.

- In the lymphoid tissues, they go through some changes to facilitate growth, and differentiation (for the performance of a specific role). This process allows them to strike at particular foreign substances as 'antibodies'.

- Natural killer cells (NK cells) focus on cancerous cells, and viral infections.

B Cells

- Also called B lymphocytes, make definite antibodies to attack particular antigens/foreign substances.

- They target viruses and bacteria dispersed in blood. The moment a B cell comes in contact with the specific irritant it was made to attack, it immediately gets altered into a plasma cell.

- B cells form plasma cells which produce a high antibody content, all of which were made for specific pathogens.

T Cells

- Also called T lymphocytes, originate from the

thymus resulting in the initial 'T' name. They are the center of what is referred to as cell-mediated immunity.

- **Cytokines** (SIGH-toh-kyens) operate as signals between cells to initiate an immune response. They are secreted majorly by the T- cells and include proteins like interferons and interleukins.

- **Interferons** (in-ter-FEAR-onz), IFNs; When viruses or cancerous cells invade the cells, the body reacts by secreting IFNs that not only triggers the immune system, and attacks the foreign substances' replication process, but also alerts neighboring cells on security consciousness.

- **Interleukins** (in-ter-LOO-kinz) are multi-functional, one of which is their role in supervising the splitting and multiplication of the B and T cells.

Additional Structures of the Lymphatic System

All the other structures comprise of lymphoid (lymph-like) tissue, and they function together with the immune system.

The Tonsils

Tonsils (TON-sils) are located behind the upper throat and nose, and made up of a defensive ring of 3 bundles of lymphoid tissue. They are essential in the immune system because they inhibit the invasion of microbes via the nose and mouth into the respiratory system.

- The **nasopharyngeal** tonsils called **adenoids** (AD-eh-noids) are found in the upper pharynx (nasopharynx)

- The **palatine** (PAL-ah-tine) (relating to the palates that form the roof of the mouth) tonsils are visible on both sides of the throat.

- The **lingual** (LING-gwal) tonsils are found at the root of the tongue, though not easily visible.

The Thymus Gland

The thymus (THIGH-mus) which is found just above the heart, progressively grows till its peak at puberty, then decreases in size as one grows older

- It triggers the growth of lymphocytes to become T cells with the help of a hormone it produces

- When mature, the T cells exit the thymus through the circulatory system.

The Vermiform Appendix

Usually called **appendix**, it juts down from the lower cecum (the initial part of the large intestine). Modern studies have linked the appendix as an accessory to the immune system even though its function had been unidentified up till recently.

The Spleen

Found in the left upper quadrant of the abdomen, behind the stomach, and under the diaphragm, is the bag-like spleen.

- It acts as a filter for blood.

- It produces leukocytes (white blood cells); Lymphocytes and monocytes that are vital components of the immune system.

- It acts as a **hemolytic recycler** (hee-moh-LIT-ick) by killing weak red blood cells and dispensing their hemoglobin to be used again. **Hemolytic** means *blood-killing*

- It also balances the quantity of erythrocytes (red blood cells) and plasma in blood by sometimes keeping erythrocytes.

The Immune System

The functions of the immune system can be compressed into one; protection of the body from unsafe matter in order to sustain and promote good health. These noxious substances include;

- Microbes which can result in illnesses

- Irritants called allergens; instigate allergies

- Toxins; noxious substances

- Malignant cells; cancerous cells that could be fatal.

If the immune system fails to prevent the invasion of any of these substances into the body, then it devises ways to get rid of them. HOW?

- Via the body's intricate signal system, the immune system can recognize, strike and memorize antigens for recall the next time it recurs.

- The system works on specificity thereby making a distinction between the body's tissues and foreign substances to prevent striking itself.

- The system's ability to memorize helps the body to prepare better for future invasion.

The Immune System's First Line of Defense

The immune system is a unique system because contrary to other body systems, it comprises of organs and vessels across various systems in the body, and isn't limited to one system. This is the first line of defenses is as follows;

- Unbroken/uncut/undamaged skin stops foreign substances from invading the body and creates an unconducive habitat for microbes as a result of its acidic wrap.

- Foreign substances that try to invade the body through breathing get trapped by nose hairs and the mucous lining of the respiratory system. Those that escape, get apprehended by the tonsils at the throats opening, or by coughing or sneezing to forcefully remove the pathogens from the body.

- Ingested food housing pathogens get digested in the stomach exposing the pathogens to the acid and enzymes secreted within.

- Leucocytes and components of the lymphatic system co-operate to rid the body of invasive substances in a specified manner.

The Antigen-Antibody Reaction

It is also called **immune reaction**, a situation where an antibody gets attached to an antigen. This reaction identifies the dangerous antigen for recognition and attack by other immunological cells.

- An **antigen** (AN-tih-jen) is a foreign or perceived foreign substance. Examples; bacteria, toxins, viruses, transplanted tissues. They stimulate an immunological response by the body.

- The attained indifference or non-reaction to a particular antigen is called **tolerance**. It also applies to drug use whereby the effectiveness of a drug reduces as a result of continuous use.

- An **antibody** (AN-tih-bod-ee) or **immunoglobulin** is a product of the immune system. It is created in reaction to the identification of a particular antigen to fight disease.

Immunoglobulins

Plasma cells produce immunoglobulins (im-you-noh-GLOB-you-lins) of which there are 5 major kinds.

Phagocytes

Phagocytes (FAG-oh-sights) are a special group of leucocytes (monocytes, macrophages, dendritic cells, and mast cells) that kill foreign substances like debris, pollen, dust, and microbes by phagocytosis. Phagocytosis, which means *to swallow a cell,* is the killing of microbes by trapping and eating them.

- **Macrophage** (MACK-roh-fayj) is a leucocyte that kills invaders by swallowing/eating. It also removes the dead invaders and warns other immunological cells to respond.

- **Dendritic cells** (den-DRIT-ic) are unique leucocytes that act as vigilantes, finding and killing disease-producing cells, and then stimulating B and T cells to action.

- **Monocytes** are leucocytes that shield the body from numerous pathogens. They form macrophages and replace both macrophages and dendritic cells.

The Complement System

There is a usual dispersion of an inactive protein group in the blood called the **complement** (KOM-pleh-ment) system. They complete or support the function of antibodies in killing pathogens.

Immunity

This is basically a state of being impervious to certain diseases. There are 2 types;

- **Natural or passive immunity** is either transferred down from mother to child during breastfeeding or existent at birth. It has nothing to do with exposure or antigen administration.

- **Acquired immunity** is gotten by previous exposure and infection by a disease.

- **Vaccination** prevents contraction of specific diseases like influenza and mumps.

- **Vaccines** are synthesized to comprise of weakened or destroyed pathogens as a defense from sickness. Illnesses like tetanus need to be boosted periodically to sustain the immunity.

Chapter 11: The Respiratory System

The roles of the respiratory system include:

- Air transport to the lungs

- Oxygen supply to the blood from inhaled air

- Exhalation to remove carbon dioxide and water (end products of metabolism) delivered to the lungs.

- Generation of laryngeal airflow that is necessary for speech.

Structures of the Respiratory System

Oxygen is delivered to the blood from the respiratory system, and supplied to all the body cells. Body cells need oxygen to survive. The respiratory system is classified into upper and lower respiratory tracts.

- **The upper respiratory tract** is made up of the trachea, larynx, pharynx, mouth, and epiglottis.

- The **lower respiratory tract** includes the lungs and bronchial tree. They are enclosed and protected by the ribcage, or thoracic cavity (thoh-RAS-ick).

- The bronchial tree and the upper respiratory tract constitute the airway.

The Nose

The nose allows air passage into the body through the nasal cavity.

- The outermost nasal opening is the **nostrils**.

- The **nasal septum** (NAY-zal SEP-tum) is a cartilaginous wall that separates the nose equally into two.

- **Cilia** (SIL-ee-ah), are the tiny hair-like structure inside the nostrils that carry out the filtration of inhaled air.

- **Mucous membranes** (MYOU-kus) form the inner lining of the nose. They also line the reproductive, urinary, and digestive systems.

- **Mucus** (MYOU-kus) is a slimy product secreted by the mucous membrane responsible for the lubrication and protection of tissues. Nasal mucus warms, moistens, and filters inhaled air.

- **Olfactory receptors** (ol-FACK-toh-ree) are nerve endings that receives smell and taste stimuli. They are found in the mucous membrane lining the top part of the nasal cavity.

The Tonsils

The lymphatic system includes the tonsils and adenoids. They defend the body against infection entry from the mouth or nose. The nasopharyngeal tonsils or adenoids are found at the roof of the mouth and behind the nose. The palatine tonsils, or tonsils are found at the end of the mouth.

The Paranasal Sinuses

Paranasal sinuses (para- near, nas- nose, -al related) are cavities lined by mucous membrane and filled with air that are found in the skull bones. A **sinus** is a cavity or sac in a tissue or organ.

Sinuses are responsible for mucus production to lubricate the tissues in the nose, making the skull bones lighter, producing sound by increasing the resonance of the voice. Some short ducts connect the sinuses to the nasal cavity.

There are 4 pairs of paranasal sinuses are named according to the bones they are found in, on both sides of the nose.

- **Frontal sinuses** are found in the frontal bone above the eyebrows. Infections in this area cause unbearable pain.

- **Ethmoid sinuses** in the ethmoid bones that are between the nose and the eyes are asymmetrical air cells. They are separated from the eye cavity by a thin layer of bone.

- **Maxillary sinuses** are the biggest paranasal sinus, and are found in the maxillary bone below the eyes. Infections in this area result in pain in the posterior maxillary teeth.

- **Sphenoid sinuses** are found in the sphenoid bone, beneath the pituitary gland, at the back of the eye and near the optic nerve. Infection in this area can cause loss of vision.

The Pharynx (Throat)

The **pharynx** (FAR-inks) takes delivery of air and food from the nose or mouth respectively.

There are 3 parts:

- **Nasopharynx** (nay-zoh-FAR-inks) 'nas- nose, '-

pahrynx'- throat. This is the first part. It's located at the back of the nasal cavity and extends behind the mouth. It exclusively transports air and continues as the oropharynx.

- **Oropharynx** (oh-roh-FAR-inks) is the second part, and it can be seen when the mouth is open. 'oro'- mouth. The oropharynx is a part of the digestive and respiratory systems as it transports food, fluid, and air to the laryngopharynx.

- **Laryngopharynx** (lah-ring-goh-FAR-inks) is the third part. It is also a part of the digestive and respiratory system. laryngo- larynx. Food, fluid moves down to the opening of the esophagus while air moves down to the opening of the trachea.

Larynx (voice box)

This is a triangular chamber found between the pharynx and trachea. Nine distinct cartilages support and protect the larynx.

- The biggest cartilage is the **thyroid cartilage** that protrudes from the throat when it is enlarged, to form the Adam's apple.

- The vocal cords are found in the larynx. The cords relax during breathing to allow air passage.

They contract when speaking and sound is generated when air is expelled from the lungs, resulting in a vibration of the cords.

Protective Swallowing Mechanisms

A portion of the pharynx is shared by the respiratory and digestive systems. While swallowing, it is possible that food or fluids go into the trachea instead of the esophagus, resulting in aspiration pneumonia or a blocked airway. This is prevented by 2 protective mechanisms that function automatically to make sure that the lungs receive only air:

- The **epiglottis** (ep-ih-GLOT-is) is a lid-shaped structure found at the base of the tongue that moves downwards and blocks the laryngopharynx to prevent food entry into the lungs and trachea

- Simultaneously, the **soft palate** stops the entry of fluid and food into the nose. It is the muscular posterior part of the roof of the mouth. It moves upwards and backwards when swallowing to block the nasopharynx.

Trachea (or windpipe)

The **trachea** (TRAY-kee-ah) carries air to the lungs. It is found anterior to the esophagus. It is kept open by a number of flexible C-shaped cartilaginous rings that facilitate the compression of food by the trachea, to allow food entry into the esophagus.

Bronchi

These are two big tubular structures known as primary bronchi (BRONG-kee). They are a continuation of the trachea, and they transport air into the two lungs. They are called the bronchial tree because they resemble an inverted tree. Singular- bronchus (BRONG-kus). There is a division of each primary bronchus inside the lungs to form progressively smaller bronchioles (BRONG-kee-ohlz).

Alveoli (air sacs)

The alveoli (al-VEE-oh-lye) are tiny, grape-like bunch located at the terminal of each bronchiole. Singular-alveolus (al-VEE-oh-lus). Carbon dioxide and oxygen are exchanged in the alveoli. Millions of alveoli are found in each lung.

- **Breathing** involves air entry from the bronchioles into the alveoli.

- The alveoli are surrounded by a system of microscopic pulmonary capillaries. **Pulmonary** (PULL-mah-nair-ee) means *related to the lungs*. The thin, flexible walls of the alveoli facilitate carbon dioxide and oxygen exchange between the blood in the pulmonary capillaries and the air in the alveoli.

- The surface tension in the lung is reduced by **surfactants**, is a detergent-like material secreted by the alveoli. Surfactants stabilize the alveoli, preventing their collapse when a person exhales.

The Lungs

They are important respiratory organs, segmented into lobes. A **lobe** is a section of an organ. The **left lung** contains two lobes, the upper and lower lobes, with the heart occupying the space in-between. The **right lung** is bigger and has three lobes, the upper, middle, and lower lobes.

Mediastinum

The central part of the chest cavity, between the lungs is called the **mediastinum** (mee-dee-as-TYE-num). It contains connective tissue and organs like the trachea, esophagus, thymus gland, lymph nodes, heart, and its arteries and veins.

Pleura

This is a thin, slimy and wet membrane that encloses the external lung surface and lines the interior of the thoracic cavity.

- **The Parietal (pah-RYE-eh-tal) pleura** forms the external pleural layer. Parietal means *pertaining to the walls of a cavity*. It encloses the diaphragm, lines the walls of the thoracic cavity, and forms a sac that houses each lung. It is attached to the chest wall.

- **The Visceral (VIS-er-al) pleura** form the inner pleural layer that encloses and forms a direct attachment with each lung. Visceral means *pertaining to internal organs*.

- **The Pleural cavity** or **pleural space** is the thin space between the visceral and pleural membrane usually filled with fluid that lubricates the membranes, enabling them to move over each other when breathing.

Diaphragm

The **thoracic diaphragm** (DYE-ah-fram) is a dome-shaped muscle layer that demarcates the thoracic cavity and the abdomen. Respiration involves the relaxation and contraction of the diaphragm. The **phrenic (FREN-ick) nerve** is responsible for the contraction of the diaphragm.

Respiration (Breathing)

This is the exchange of oxygen for carbon dioxide. One breath or respiration is made of one inhalation and one exhalation. Air movement into and out of the lungs is also called **ventilation**.

Inhalation and Exhalation

The process by which the diaphragm contracts and moves downwards to take in air is called **Inhalation** (in-hah-LAY-shun). Here, there is an expansion of the thoracic cavity and this creates a space that facilitates air entry into the lungs.

The process of breathing out is called **exhalation** (ecks-hah-LAY-shun). Relaxation of the diaphragm causes it to move upwards, making the thoracic cavity become narrower so that air can be expelled from the lungs.

External Respiration

This is the process of taking air in and out of the lungs from the external environment while simultaneously exchanging oxygen for carbon dioxide. Inhalation of air into the alveoli allows oxygen entry into the capillaries, and its transport by the red blood cells to other body cells. Concurrently, carbon dioxide in the bloodstream is delivered to the lungs for exhalation.

Internal Respiration

Also called **cellular respiration**, this is the gaseous exchange that takes place in the cells of the blood and tissues.

Oxygen is delivered to the body cells from the blood, and the cells generate carbon dioxide which is transported into the blood. The blood then delivers the carbon dioxide to the lungs where it is exhaled.

Chapter 12: The Digestive System

The main component of the digestive system is the **gastrointestinal tract** (gas-troh-in-TESS-tih-nal) or **GI tract**. Gastro means *stomach*, intestin means *intestine*, al means *related to*. The organs work together with accessory organs (liver, pancreas, and gall bladder)

- The **upper gastrointestinal tract** includes the stomach, esophagus, throat (pharynx), and mouth. It allows food entry into the body before the stomach initiates digestion.

- The **lower gastrointestinal tract** or the bowels include the small intestine, large intestine, rectum, and anus. Digestion is completed here and waste is prepared for excretion.

The Oral Cavity (Mouth)

The organs in the oral cavity include the lips, teeth, soft and hard palate, tongue, periodontium, and salivary glands

Lips

The **Lips** or **labia** (Labium-singular) surround the opening of the mouth. Labia also refer to parts of the female genitals. When feeding, food is held in the mouth by the cheeks, tongue, and lips. The lips are also essential in respiration, speech, and display of emotions.

Palate

The roof of the mouth is formed by the palate (PAL-at). It is important in producing speech sounds, and snoring. It has 3 divisions:

- **Hard palate** forms the anterior part of the palate. It is covered with specialized mucous membranes which consist of irregular folds called **rugae** (Ruga-singular).

- **Soft palate** is the elastic, posterior part of the palate. It blocks the nasal cavity when swallowing to stop the upward movement of food and fluid into the nasal cavity.

- **Uvula** (YOU-view-lah) dangles from the edge of the soft palate. It moves up with the soft palate when swallowing.

Tongue

The tongue is muscular and strong. The anterior part is mobile and flexible, and the posterior part is fixed. The tongue is essential to speech, chewing, and swallowing.

- The **dorsum** is the upper part of the tongue. It has a hard protective covering and contains tiny bumps called **papillae** (pah-PILL-ee) in some areas. **Taste buds** are the receptors for taste, and are located in the papillae.

- The sublingual tongue surface and the tissues beneath the tongue are covered with soft, vascularized tissues. Sublingual means *below the tongue*. Vascularized means *plenty blood vessels*. The abundance of the blood vessels makes it possible to administer some drugs sublingually by placing them below the tongue where they are rapidly absorbed into the blood.

Soft Tissues of the Oral Cavity

- **Periodontium** (per-ee-oh-DON-shee-um) refers to the structures that support and surround the teeth. Peri means *around*, odonti means *teeth*, -um is a suffix. It includes the dental arches and soft tissues.

- **Gingiva** (JIN-jih-vah) or **gums**, or **masticatory mucosa** is the specialized mucous membrane around the neck of the teeth. It encloses the dental arches. Gingivae - plural.

Dental Arches

These are the bony structures in the mouth. They stabilize the teeth, and enable speech and chewing.

- **Temporomandibular** (tem-poh-roh-man-DIB-you-lar) or **TMJ** is formed at the posterior part of the mouth by the union of the mandibular and maxillary arches.

- The **maxillary arch** or **upper jaw** is fixed. It includes the bones of the lower surface of the skull.

- The **mandibular arch** or **lower jaw** is a distinct bone. It is the only mobile bone in the joint.

Teeth

The arrangement of the teeth in the upper and lower jaws is called **dentition** (den-TISH-un). There are 4 classes of teeth in humans:

- **Cuspids**: canines and incisors, are used to tear and bite.

- **Bicuspids**: molars and premolars are used in grinding and chewing.

Primary and Permanent Dentition

- **Primary dentition,** or **baby teeth,** or **deciduous dentition** are the teeth that erupt at the beginning of childhood. They are 20 in number, and are substituted by the permanent teeth in late childhood. It includes four canines, eight incisors, eight molars, and zero premolars.

- **Permanent dentition** includes the 32 teeth that are present throughout life. The primary teeth are replaced by 20 of them and the remaining 12 erupt at the back of the mouth. It includes four canines, eight incisors, twelve molars, and eight premolars.

Occlusion in dentistry refers to the contact between the chewing surfaces of the lower and upper teeth.

Structures and Tissues of the Teeth

- The part of the tooth that can be seen in the mouth is the **crown**; it is enclosed by the **enamel,** the toughest material in the body.

- The teeth are anchored by their **roots** within the dental arch. The **cementum,** another hard material protects the root.

- The **neck of the tooth** or **cervix** is where the crown and root meet.

- A large part of the tooth is made of **dentin**. The **enamel** encloses the part above the gum, and the **cementum** covers the root.

- The **pulp cavity** is the space in the crown and root of the teeth that is enclosed by the dentin to protect the soft tooth pulp. The **root canal** is an extension of the pulp in the root. The abundant network of blood vessels in the pulp supplies nutrients and innervation to the tooth.

Saliva and Salivary Glands

The mouth is kept wet by **saliva**, a colorless fluid. It keeps the teeth healthy and initiates digestion by lubricating food when swallowing and chewing.

Saliva is secreted by 3 pairs of salivary (SAL-ih-ver-ee) glands and is transported to the mouth via ducts.

- **The parotid glands** are found on the face, anterior to each ear. The parotid gland ducts are found beside the upper molars, in the interior of the cheeks.

- **The sublingual glands** and ducts are found beneath the tongue, on the floor of the mouth

- **The submandibular gland** and ducts are found close to the mandible, on the floor of the mouth.

Pharynx

The pharynx provides passage for digestion and breathing. It is also particularly essential to deglutition or swallowing. When swallowing, food is prevented from entering into the pharynx by the closure of the trachea by the **epiglottis** (ep-ih-GLOT-is), a lid-shaped structure.

Esophagus

The muscular tube that carries ingested food from the pharynx to the stomach is the **esophagus** (eh-SOF-ah-gus). The **cardiac sphincter** or **lower esophageal sphincter** is a ring-shaped muscle between the stomach and esophagus. Food entry into the stomach is facilitated by its relaxation when swallowing. The backward flow (**regurgitation**) of the stomach content is prevented by the closure of this sphincter.

Stomach

This is a sac-shaped structure divided into the fundus (top, rounded part), body (major part), and antrum (lower section).

- **Rugae** (ROO-gay) are the ridges in the mucosa membrane in the stomach. The alteration in their size makes the stomach flexible. Gastric juice is produced by glands between these ridges. The initial stage of digestion is facilitated by gastric juices. They produce mucus which forms a protective layer over the stomach lining.

- **Pyloric sphincter** (pye-LOR-ick) is the muscular ring at the base of the stomach which controls the movement of undigested food to the small intestine from the stomach.

- **Pylorus** (pye-LOR-us) is the narrow passage that connects the stomach to the small intestine.

Small Intestine

It continues from the pyloric sphincter till the first section of the large intestine. It is 20 feet long and divided into three parts where digestion of food and absorption of nutrients into the blood takes place.

- **Duodenum** (dew-oh-DEE-num) is the first part of the small intestine. It continues from the pylorus to the jejunum.

- **Jejunum** (jeh-JOO-num) is the second part of the small intestine extending from the duodenum to the ileum.

- **Ileum** (ILL-ee-um) is the longest part of the small intestine, extending from the jejunum to the large intestine.

Large Intestine

It continues from the end of the small intestine till the anus. It is two times the width of and ¼ times the length of the small intestine. Waste produced after digestion is prepared for elimination via the anus. The cecum, colon, and anus constitute the main parts of the large intestine.

Cecum

Positioned on the right side of the abdomen, the **cecum** (SEE-kum) is a sac that continues from the end of the ileum to the start of the colon.

- The flow of content from the ileum to the cecum is controlled by the **ileocecal** (ill-ee-oh-SEE-kull) sphincter.

- The **appendix**, or **veriform appendix**, dangles from the lower part of the cecum, and it contains lymphoid tissue. Veriform means *worm-shaped*.

The Colon

This is the longest section of the large intestine, it has four divisions:

- **The ascending colon** that moves up from the cecum to the base of the liver.

- **The transverse colon** that crosses the right side of the abdomen to the left, close to the spleen.

- **The descending colon** moves downward from the left side of the abdomen to the sigmoid colon.

- **The sigmoid colon** (SIG-moid), an S-shaped structure that extends from the descending colon above to the rectum.

The Rectum and Anus

The broadest part of the large intestine is the **rectum**. It constitutes the final four inches of the large intestine and stops at the anus.

The opening of the anus marks the end of the GI tract.

The **external anal sphincter** and **internal anal sphincter** directs the flow of waste through the anus.

Anorectal (ah-noh-RECK-tal) is a combination of the anus and rectum; ano means *anus*, rect means *rectum*, and -al means *related*.

Accessory Digestive Organs

Organs that are not components of the digestive system but perform a crucial function in the process of digestion are called **accessory digestive organs**. They include pancreas, gallbladder, and liver.

Liver

The largest body organ is the liver. It is essential to the removal of toxins from the blood and changing food into nutrients and energy required by the body. Hepatic means *related to the liver*; 'hepat'- liver, '-ic'- related.

- Excess glucose is eliminated from the blood by the liver and stored as glycogen. If there is a drop in the level of blood sugar, glycogen is changed back to glucose by the liver and released to the body.

- The digestion of fat is aided by **bilirubin** (bill-ih-ROO-bin) or **bile**, a yellow-green liquid produced by the liver. It is then transported to the gallbladder where it is stored and concentrated. In large quantities, bilirubin in the body can cause jaundice and other diseases.

Biliary Tree

The transport of bile to the small intestine from the liver is facilitated by the **biliary tree** (BILL-ee-air-ee). Biliary means *related to bile*.

- The fusion of many small ducts in the liver forms the biliary tree, with the trunk called the **common hepatic duct** lying external to the liver.

- The common hepatic duct transports bile from the liver to the gall bladder. Bile however leaves the gall bladder via the cystic duct.

- After exiting the gall bladder, the combination of the cystic duct and the common hepatic forms the **common bile duct**. The pancreatic duct and common bile duct then enters the duodenum.

Gallbladder

This is a pear-shaped organ as big as an egg, located under the liver. It handles the storage and concentration of bile. The contraction of the gallbladder ejects bile from the biliary tree. Cholecystic (koh-luh-SIS-tick)-

related to the gallbladder; 'cholecyst'- gallbladder, '-ic'- related to

Pancreas

Pancreas (PAN-kree-as) is a tender, elongated gland, six inches in length, located at the back of the stomach. It is essential to the digestive and endocrine systems. The manufacture and secretion of pancreatic juice is done by the **pancreas**. The **pancreatic duct** transports pancreatic juice outside the pancreas and meets the common bile duct prior to its entry into the duodenum.

Digestion

The process of breaking down complex foods to form nutrients that can be easily used by the body is called **digestion**.

- **Digestive enzymes** carry out the chemical process of breaking down food into smaller forms of nutrients for the body's use.

- **Nutrients** are substances derived from food. They are essential for the body to function properly. **Primary nutrients** include protein, carbohydrates, and fats.

- **Essential nutrients** are needed in only small quantities and they include minerals and vitamins.

Metabolism

Metabolism (meh-TAB-oh-lizm) refers to all the processes involved in the utilization of nutrient by the body; metabolic means *change*, ism means *state*. It involves two parts which are the reverse of each other:

- **Catabolism** (kah-TAB-oh-lizm) is the breakdown of cells and substances to produce carbon dioxide and energy.

- **Anabolism** (an-NAB-oh-lizm) is the buildup of cells or substances from nutrients.

Absorption

The process of transporting the products of digestion to body cells is called **absorption** (ab-SORP-shun).

- The mucosa lining of the small intestine is covered with finger-shaped protrusions known as **villi** (VILL-eye). One villus consists of lacteals and blood vessels.

- Nutrients are directly absorbed from the digestive system by the blood vessels and then transported to body cells.

- Fat-soluble vitamins and fats cannot be

transported directly to the blood; rather they are absorbed by specialized lacteals which then transport them through lymphatic vessels. During the transport of the nutrients, **lymph nodes** serve as a filter before their transport to the blood.

The Role of the Mouth, Salivary Glands, and Esophagus

The process of breaking down food into smaller parts and mixing with saliva in preparation for swallowing is called **mastication** (mass-tih-KAY-shun).

- A mass of chewed food ready for swallowing is called **bolus** (BOH-lus). Bolus could also mean a route of drug administration. Swallowing involves the movement of food from the mouth, to the pharynx, then to the esophagus.

- **Peristalsis** and gravity propels food downwards within the esophagus. **Peristalsis** (pehr-ih-STAL-sis) is a number of unidirectional wave-like contractions of the smooth muscles that propels food into the digestive system.

The Role of the Stomach

The process of digestion is initiated by the digestive enzyme and hydrochloric acid contained in the gastric juices of the stomach. The stomach walls allow the exit of some nutrients into the bloodstream.

Food is converted into chyme by gastric juice and the churning action of the stomach. **Chyme** (KYM) is the partially liquid mass of incompletely digested food which exits the stomach into the small intestine via the **pyloric sphincter**.

The Role of the Small Intestine

- Peristalsis propels food in the small intestine and this begins the process of converting food into useful nutrients.

- Within the duodenum, chyme is combined with bile and pancreatic juice.

- **Emulsification** is the process by which bile degrades big fat globules so that it can be easily digested by pancreatic enzymes.

- It is essential to the absorption of nutrients. Jejunum secretes large quantities of digestive enzymes that enable the continuation of digestion.

- Ileum plays a key role in nutrient absorption from digested food.

The Role of the Large Intestine

The large intestine collects the end products of digestion and stores them before the body excretes them.

- Waste products enter the large intestine in fluid form. The walls of the large intestine reabsorbs excess water, maintaining the fluid balance of the body. Feces are formed from the remaining food waste. **Feces** (FEE-seez) or solid body wastes are excreted from the rectum and anus.

- The process of clearing the large intestine is known as **bowel movement** (BM) or **defecation** (def-eh-KAY-shun).

- Billions of bacteria are found in the large intestine; most are harmless and they help in degrading organic waste in a process that emits gas.

- Gas movements in the stomach produce a rumbling noise called **borborygmus** (bor-boh-RIG-mus).

- The passage of gas through the rectum is known as **flatulence** (FLAT-you-lens) or **flatus**.

Chapter 13: The Reproductive System

The reproductive organs and their associated structures are referred to as the **genitalia** (jen-ih-TAY-lee-ah). The genitals are divided into to:

- The internal genitalia (reproductive organs located within the body cavity).

- The external genitalia (reproductive organs located outside the body).

The **perineum** (pehr-ih-NEE-um) of both males and females is located between the coccyx and the pubic symphysis. It serves as the external surface region of the genitals. The tissue of the male perineum extends from the scrotum downwards, while those of females extend from the pubic symphysis towards the anus.

The Male Reproductive System

The male reproductive system is designed to produce and deliver sperm into the female body. When the sperm cells fertilize an ovum (egg), a new life is created.

The male external male genitalia are:

- The scrotum which encloses two testicles.

- The penis

The internal male genitalia includes all other structures related to the male reproductive system.

The Scrotum and Testicles

The sac-like structure that protects, and supports the testicles is known as the **scrotum** (SKROH-tum). It can be found suspended between the pubic arch behind the thighs.

- Within these sacs are found the two egg-shaped glands known as the **testicles** or **testes**. Sperm production takes place within the **seminiferous tubules** (see-mih-NIF-er-us TOO-byouls) of the testes.

- Located at the proximal part of each testicle is the **epididymis** (ep-ih-DID-ih-mis) which serves as sperm reservoirs. The tube progressively narrows to form the vas deferens.

Semen Formation

- **Sperm** (spermatozoa) are male gametes. At the peak of male sexual excitement, the **semen**

(SEE-men) (the fluid containing sperm) is ejaculated into the vagina.

- The entire process of sperm formation and maturation is known as **spermatogenesis** (sper-mah-toh-JEN-eh-sis).

- The scrotum regulates the temperature necessary for optimal sperm production (93.2°F) by adjusting the distance between the body and the testicles.

- After sperm production, the sperm cells move upward from the epididymis into the body and then into the vas deferens where the prostate gland and the seminal vesicles secrets seminal fluids to form semen.

The Penis

The male external sexual organ is the **penis** (PEE-nis). The penis transports and delivers sperm into the vagina.

- The **erectile tissues** are responsible for erection. When sexual stimulated, these tissues becomes filled with blood which makes the penis hard and stiff.

- The head of the penis, the **glans penis** (glanz PEE-nis), is a sensitive region. The glans penis is covered by a retractable foreskin known as the **prepuce**.

The Vas Deferens, Seminal Vesicles, and the Ejaculatory Duct

The Vas Deferens

The **vas deferens** (vas DEF-er-enz) or **ductus deferens** are the long distal ends of each epididymis that is connected to the urethra.

The seminal vesicles

The **seminal vesicles** (SEM-ih-nal) open into the vas deferens as it joins the urethra. These glands secrete a thick, yellowish fluid which nourishes the sperm cells. Seminal secretion forms about 60% of the total volume of semen.

The ejaculatory duct

The **ejaculatory duct** begins at the vas deferens, passes through the prostate gland, and finally drains into the urethra. During ejaculation, reflex action causes the semen to be ejected out of the penis through the urethra.

The Prostate Gland

Right under the bladder is located the **prostate gland** (PROS-tayt). The prostate gland secretes a thick, alkaline fluid during ejaculation which aids sperm motility.

The Bulbourethral Glands

The **bulbourethral glands** (bul-boh-you-REE-thral) or **Cowper's glands** can be found right below the prostate gland. Each of the glands is located on one side of the urethra and is connected to the urethra.

During sexual stimulation, these glands secrete fluids that lubricates the urethra for easy passage of the sperm.

The Urethra

The male urethra serves both urinary and reproductive functions. The urethra extends into the penis and leads out of the body.

The Female Reproductive System

The female reproductive system is basically designed for reproduction. The female reproductive organs include the ovaries, the uterus and the breast among others.

The organs of the female reproductive system are also divided into external and internal genitalia.

The external female genitalia:

- The external female genitalia are collectively known as the **vulva** (VUL-vah) or the **pudendum,** and it is located behind the **mons pubis** (monz PYOU-bis). The vulva consists of the vaginal orifice, clitoris, labia, and the Bartholin's glands.

- The **vaginal lips** are known as the **labia minora** and **labia majora** (LAY-bee-ah mih-NOR-ah and LAY-bee-ah mah-JOR-ah) and they serve as a protective cover to the other external genitalia including the **urethral meatus**.

- The **clitoris** (KLIT-oh-ris) like the glans penis is an erectile and highly sensitive organ located anteriorly to the vaginal orifice.

- The **Bartholin's glands** are two small glands that secrets mucus which lubricates the vagina and they are located on both sides of the vaginal orifice. On the surface of the vaginal orifice is a mucous membrane that partially covers this opening known as the **hymen** (HIGH-men). This membrane can be torn when subjected to stressful activities or during the first instance of intercourse.

The breasts

- The breasts are made up of connective tissue, the mammary glands and fat (mamm/o and mast/o both mean breast). **Suspensory ligaments** attaches each breast to the pectoral muscles.

- The mammary glands or lactiferous glands develops during puberty and are responsible for milk-producing. The milk produced are channeled from the mammary glands to the nipple by the **lactiferous ducts** (lack-TIF-er-us). From the nipples, breast milk flows to the **areola** (ah-REE-oh-lah), a dark-pigmented area.

The Internal Female Genitalia

Most of the internal female genitalia are deeply seated within the pelvic cavity and well protected by the pelvis bone. They include the vagina, the uterus, the fallopian tubes, and the ovaries.

Vagina

The word parts colp/o and vagin/o both mean **vagina**. The vagina (vah-JIGH-nah) is a muscular tube with a mucosa layer extending from the cervix and opening to the outside of the body.

The Uterus

- The **uterus** (YOU-ter-us), or the **womb** is located midway between the pubic and the sacral bones, and between the rectum and the urinary bladder.

- The womb is a pear-shaped organ with muscular walls and a mucous membrane lining that is richly supplied with blood by the blood vessels.

- The uterus serves as the home of the growing fetus after fertilization, and is bent forward in a position known as **anteflexion** (an-tee-FLECK-shun).

The Fallopian Tubes

- There are two **uterine tubes** or **fallopian tubes** (fal-LOH-pee-an). These tubes serve as a connection between the ovary and the upper end of the uterus.

- The funnel-shaped opening at the distal end of the fallopian near the ovary is called the **infundibulum** (in-fun-DIB-you-lum).

- The finger-like extensions at the proximal end of the fallopian tube are known as the **fimbriae** (FIM-bree-ee). The fimbriae play the major role of catching the mature ovum released by the ovary.

- Every monthly, the fallopian tubes transport one mature ovum from the ovary to the uterus. Also, after sexual intercourse, the tubes carry sperm upward from the uterus toward the descending matured ovum to aid fertilization.

The Ovaries

- Both ovaries (OH-vah-rees) are located in the lower abdomen, each on either side of the uterus.

- The **ovaries** are a pair of almond-shaped organs that contains thousands of **follicles** (FOL-lick-kuls).

- A follicle is a fluid-filled sac containing a single ovum (egg).

- The ovaries also secrete the sex hormones progesterone and estrogen.

- Each month (after puberty) an ovum matures and is released.

Menstruation

The periodic sloughing off of the endometrial lining along with unfertilized egg from uterus is known as menstruation (men-stroo-AY-shun) or menses. The first experience menstruation usually begins at puberty and it is called menarche (MEN-ar-kee). The age at which an individual starts menstruating differs from person to person but the United States, the average age is 12 (the average menstrual cycle consists of 28 days).

The physiological termination of the menstrual function in a woman is termed menopause (MEN-oh-pawz) and it usually occurs during the middle age. Menopause can only be confirmed when a woman had missed her period for at least 1 year.

The transition phase between regular menstrual periods and no periods at all is termed perimenopause (pehr-ih-MEN-oh-pawz) and this can last as long as 10 years. During this phase, hormonal imbalance can cause symptoms, such as hot flashes,irregular menstrual cycles,disturbed sleep, and mood swings.

Terms Related to Pregnancy and Childbirth

Ovulation

The release of matured egg from a follicle is known as **ovulation** (ov-you-LAY-shun). Ovulation occurs approximately every the 13th or 14th day of a woman's menstrual cycle. Once the ovum is released, the fimbriae of the fallopian tube catches it and in wave-like peristaltic actions, it is pushed downwards through the fallopian tube towards the uterus and this usually takes about 5 days to pass through the fallopian tube. Along the tube, if the ovum meets with a sperm, it will fertilize the ovum within the fallopian tube.

The ruptured follicle enlarges, to become the **corpus luteum** (KOR-pus LOO-tee-um) which secretes the hormone **progesterone**. This hormone helps to prepare the uterine lining for the implantation of fertilized egg. If fertilization thus occurs, the corpus luteum continues to secrete the progesterone which is necessary to maintain the pregnancy.

If there is no fertilization, the corpus luteum shrinks and the endometrium lining gradually sloughs off resulting in what is called the **menstrual flow**.

Fertilization

During **sexual intercourse** or **coitus** (KOH-ih-tus), about 100 million sperm cells are ejaculated into the vagina. **Conception** occurs when a sperm fertilizes an ovum. The fusion of the sperm and ovum leads to the formation of a **zygote** (ZYE-goht). The zygote then migrates towards the uterus where implantation or embedding of the zygote occurs. Starting from when implantation occurs till the 8th week of pregnancy, the developing zygote is referred to as an **embryo** (EM-bree-oh) but from the 9th weeks till birth, the developing child is called a **fetus** (fet means *unborn child*, and -us is a singular noun ending).

Multiple Births

Fertilization of more than one egg could occur if

multiple eggs passing down the fallopian tube encounter sperm cells on their way up the fallopian tube.

Identical twins result from the division of a single egg cell that has been fertilized by a single sperm cell to form two embryos. Thus, each of the twins inherits the same genetic information from their parents.

Fraternal twins on the other hand, result from the fertilization of individual ova by different sperm cells thus developing into two different embryos.

The Chorion and Placenta

The thin outer membrane that encloses the embryo is known as the **chorion** (KOR-ee-on) and it forms part of the placenta.

The **placenta** (plah-SEN-tah) serves as a temporary organ which prevents the maternal blood and fetal blood from mixing, yet it allows the exchange of oxygen, nutrients, and waste products between the fetus and the mother. In addition, the placental barrier allows drugs and other chemicals to pass through as well.

The placenta produces certain hormones which are necessary to maintain the pregnancy. But after delivery, the placenta and fetal membranes are expelled.

The Amniotic Sac

The innermost membrane that encloses the embryo within the uterus is the **amniotic sac** (am-nee-OT-ick) or the **amnion**. The embryo is surrounded by the amniotic cavity, and develops within the fluid-filled amniotic sac.

The fluid within the amniotic sac is known as the **amniotic fluid** (am-nee-ON-ick), and it helps to protect the fetus. It also support it's floating movement.

The Umbilical Cord

Blood, oxygen, waste, and nutrients are transported from the placenta to the developing child through a tube-like structure called the **umbilical cord** (um-BILL-ih-kal). After birth, the umbilical cord is cut off just before the placenta is delivered.

Gestation

- The **gestation** (jes-TAY-shun) or **pregnancy period** usually takes approximately 40 weeks (280 days). During this period, the fetus develops within the mother's uterus until the end of the gestation period when the fetus will be ready for birth.

- The gestation period can also be divided into three trimesters, each lasting for about 13 weeks and it begins from the first day of the last menstrual period (LMP). The period of rest after child delivery is termed **confinement**. The first recognizable movement of the fetus felt by the mother is known as **quickening,** and this usually takes place between the 16th to 20th weeks of gestation.

- As the pregnancy progresses, **Braxton Hicks contractions** (intermittent painless uterine contraction) occurs with increasing frequency. These contractions should not be confused with actual labor pains because they are usually irregular, and painless.

- The viability of a fetus is determined by its ability to live outside the uterus, and these depend on number of factors such as the birth weight, developmental stage and the developmental age of the lungs of the fetus.

- **Antepartum** (an-tee-PAHR-tum) is the final period of pregnancy prior to the onset of labor.

The Mother

- A woman who has never given birth is known as a **nulligravida** (nulli means *none*, and gravida means *pregnant*).

- A woman who has never given birth to a viable child is referred to as a **nullipara** (nulli means *none*, and para means *to bring forth*).

- A woman during her first pregnancy is called **primigravida** (primi means *first*, and gravida means *pregnant*).

- A woman with a record of at least one viable child is termed **primipara** (primi means *first*, and para means *to bring forth*).

- A woman who has given birth multiple times is called a **multiparous** (multi means *many*, and parous means *having borne one or more children*).

Childbirth

Parturition or childbirth takes place in three stages. The stages of labor and delivery are:

- Dilation of the cervix

- Delivery of the baby

- Expulsion of the placenta

The First Stage:

The first stage is usually the longest stage of labor. It involves the progressive dilation and effacement of the cervix, and the eventual rupture of the amniotic sac leading to the release of amniotic fluid (broken water). During the effacement process, the cervix gradually becomes softer and thinner in preparation for delivery.

The Second Stage:

The second stage is initiated by the dilation of the cervix (to about 10 centimeters) to facilitate delivery. At this point, the uterine contraction becomes progressively stronger and frequent. This contraction is not sufficient enough to bring about delivery hence, the mother pushes hard to help expel the child through the **birth canal** (vagina).

The Third Stage:

After childbirth, placenta expulsion follows.

Postpartum

The period after childbirth is termed **postpartum** (pohst-PAR-tum).

The Mother

Puerperium

The period after the delivery of the placenta to approximately the first six weeks is known as the **puerperium** (pyou-er-PEE-ree-um). After this period, all physiological changes in the mother's body due to pregnancy are reversed.

Lochia

After childbirth, vaginal discharge usually occurs for about 4 to 6 weeks and this is known as **lochia** (LOH-kee-ah). This discharge consists of mucus and blood.

Uterine involution

The involution of the uterus helps to return the uterus to its normal size and condition after delivery.

Colostrum

The Colostrum (kuh-LOS-trum) is the first set of milk produced by the mammary glands during late pregnancy and the first few days after giving birth. It is rich in essential nutrients and antibodies needed by the infants.

Lactation

The process of milk formation and secretion from the

breast is termed **lactation** (lack-TAY-shun). A few days after birth, the colostrum is replaced with breast milk.

Postpartum depression

After child delivery, the mother might experience some feelings of depression or mood disorder characterized by loss of pleasure in normal activities and feelings of sadness due to sudden but rapid changes in the hormonal levels. In severe cases, treatment may be required.

The Baby

During the first 4 weeks after birth, the newborn infant is known as a **neonate** (NEE-oh-nayt).

The first few stools of a neonate are formed by greenish material that collects in the intestine of a fetus known as **meconium** (meh-KOH-nee-um).

Apgar Scores

After birth (between 1 and 5 minutes) the physical status of a neonate can be evaluated using the Apgar score.

This is done by assigning numerical values from 0 to 2 to each of five criteria:

- Heart rate,

- Respiratory effort,

- Muscle tone,

- Response stimulation, and

- Skin color.

Interpretation:

A total score of 8 to 10 indicates the best possible physiological condition.

Chapter 14: Other Sensory Systems

The Eyes

The eyes are the major organ for sight and they are used to receive and send images to the brain.

Abbreviations relating to the eyes (These terms should be written out and not abbreviated when used):

- Right eye - oculus dexter (OD)

- Left eye - oculus sinister(OS)

- Both eyes or each eye - oculi uterque or oculus uterque (OU) respectively.

The singular of eye is **oculus** and the plural of eye is **oculi**

The eye consists of the eyeball as well as the adnexa which not only connects but also surrounds the eyeball

The Adnexa of the Eyes

The structures outside the eyeball is called the adnexa of the eyes and can also be called **adnexa oculi**. These structures include the orbit, eyelids, conjunctiva, eye muscles, lacrimal apparatus and eyelashes. **Adnexa** (ad-NECK-sah) is used to signify the adjoining anatomical or accessory parts of an organ and it is also in plural form.

The Orbit

The eye socket is also known as the orbit, and is the bony cavity that not only protects but also houses the eyeball and its related nerves, associated muscles, and blood vessels in the skull.

Muscles of the Eye

- There are six major eye muscles arranged in three pairs are connected to both eyes. They include the **lateral and medial rectus muscles**, the **superior and inferior oblique muscles** and the **superior and inferior rectus muscles**.

- These muscles makes it possible for a wide range of accurate eye movements to occur. **Rectus** is translated to straight while **oblique** is an angle that is slanted but neither parallel nor perpendicular.

- **Binocular vision** is gotten from bin- *two*, ocul-*eye*, and -ar *pertaining to*. This is when both eyes muscles begins to work together to make possible the normal depth perception. **Depth perception** occurs when one can see things in three different dimensions.

The Eyelids, Eyebrows, and Eyelashes

The upper and lower eyelids as well as the eyebrows and eyelashes all work together to protect the eyeball from excess light, foreign matter, and injuries arising from other causes.

- **Canthus** (KAN-thus): Canth - *corner of the eye*. Plural - canthi. Canthus is the angle where the upper and lower eyelids come together.

- **Sebaceous glands** that produce oil are present at the edges of the eyelids.

- **Cilia** (SIL-ee-ah) are the little hairs that make up the eyelashes and eyebrows. These hairs can also be found in the nose and they help to prevent foreign materials from entering the nose.

- The **tarsus** also called the **tarsal plate** (TAHR-suhs) gives the required shape and stiffness necessary as the framework of the upper and lower eyelids. Tars - *edge of the eyelid;* plural form - *tarsi.*

Note: Tarsus could also mean the seven tarsal bones present in the foot's instep.

The Conjunctiva

This is a clear mucus membrane present at the underside of both eyelids and goes on to form a protective covering over the exposed part of the eyeball. Conjunctiva (kon-junk-TYE-vah). Plural - conjunctivae.

The Lacrimal Apparatus

The **lacrimal apparatus** also called the **tear apparatus** consists of structures responsible for producing, storing and removal of tears. Lacrimal (LACK-rih-mal). The process of secreting tears is known as **lacrimation**.

- **Tears** also known as **lacrimal fluid** is secreted by the lacrimal glands which are found underneath the upper eyelid above the outer corner of both eyes.

- Maintenance of moisture on the anterior surface of the eyeball is the function of the lacrimal fluid (tears).

- The lacrimal fluid is spread across the eye through blinking.

- The passageway which drains excess tears to the nose is called the **lacrimal duct** (**nasolacrimal duct**).

- A duct at the inner corner of the eye. The function of this duct is to collect and empty tears into the lacrimal sacs. The over flowing of tears from the lacrimal canals is what is known as **crying**.

- The enlargement of the upper part of the lacrimal duct is known as the lacrimal sac or tear sac.

The Eyeball

The eye also referred to as the **globe**, is a sphere which is an inch in size although only aboutone-sixth of its surface can actually be seen.

- **Optics** (OP-tik): opt – *sight,* and -ic - *pertaining to.* Optic pertains to the eye or sight.

- **Ocular** (OCK-you-lar): ocul - *eye.* Ocular means pertaining to the eye.

- **Extraocular** (eck-strah-OCK-you-lar): Extraocular means outside the eyeball.

- **Intraocular** (in-trah-OCK-you-lar): intra -

within. Intraocular means within the eyeball.

Walls of the Eyeball

The wall of the eyeball comprises of three layers and they are:

- The **retina** (RET-ih-nah) is the most sensitive part of the eye. It is the innermost layer that lines the posterior segment of the eye. The retina receives and sends nerve impulses through the optic nerve to the brain. The **optic nerve** is also referred to as the **second cranial nerve**.

- **Choroid** (KOH-roid) is also called the is gotten from the words. It is the opaque middle layer of the eyeball which houseslots of blood vessels and also supplies blood to the entire eye. **Opaque** refers to substance that light cannot pass through.

- The **sclera** (SKLEHR-ah) is also known as the white of the eye. It helps to maintain the shape of the eye and also protects the inner layers of tissuewhich are fragile. Apart for the surface covered by the cornea, the sclera which is a fibrous and tough tissue makes up the outer layer of the eye.

197

Segments of the Eyeball

The inner segment is divided into two and they are the posterior and anterior segments.

Posterior Segment of the Eye

The posterior segment of the eye consists of the remaining two-thirds of the eyeball. It is lined with the retina and also contains the **vitreous humor**. The vitreous humor (VIT-ree-us is also known as **vitreous gel**, and is a soft, clear, jelly-like mass that is filled with millions of fine fibers. These fibers are in turn connected to the surface of the retina and help to maintain the shape of the eye.

Anterior Segment of the Eye

This segment forms one-third of the front of the eyeball. This segment is further split into the following chambers:

- The **anterior chamber** is present in found in front of the iris and behind the cornea.

- The **posterior chamber** is found in front of the ligaments holding the lens in their positions and behind the iris.

Aqueous humor is also referred to as aqueous fluid and it fills both the posterior and anterior chambers. **Aqueous** (AH-kwee-uhs) means *containing water or being* *watery,* while humor in this case means *a substance that is semi fluid or any body of liquid that is clear.*

The aqueous humor not only gives nourishment to the intraocular structures but also helps to maintain the shape of the eye. The aqueous humor is filtered regularly and drained through both the trabecular meshwork and the **canal of Schlemm**.

Intraocular pressure (IOP) is used to measure the pressure of the fluid in the eye. The rate at which the aqueous humor comes into the eye and leaves the eye is used to regulate the intraocular pressure.

Structures of the Retina

- **The rods** are black and white receptors while the **cones** are color receptors. They are both present in the retina and are used receive images that have passed through the eye's lens. These images are then changed to nerve impulses and are sent to the brain through the optic nerve.

- **Choroid** (KOH-roid) is also called the **choroid coat**. It is the opaque middle layer of the eyeball which houses lots of blood vessels and also supplies blood to the entire eye. **Opaque** means *substance that light cannot pass through.*

- The **macula** (MACK-you-lah) is also called the

macula lutea. It is the light-sensitive area at the centre of the retina which can be seen clearly and it is what makes sharp central vision possible.

Note: Macula also means a small spot. A macula also referred to as a macule, could also mean a small, discolored spot on the skin, for example a freckle.

- The **fovea centralis** (FOH-vee-ah sen-TRAH-lis) is a pit at the middle of the macula. This area is filled with cones but no rods and this makes it give the best color vision.

- The **optic disk** is a small area in the eye where the nerve endings of the retina goes into the optic nerve. It is also known as the **blind spot** because there are no cones or rods present in it to change images to nerve impulses.

- The optic nerve sends these nerve impulses to the brain from the retina.

The Uvea

The **uvea** (YOU-vee-ah) is the layer of the eye that is pigmented. It consists of iris, ciliary body and choriod. It also has a rich blood supply.

The Iris

- The **iris** is a circular structure which surrounds the pupil and it is also colorful. The amount of light which is permitted to enter the eye through the pupil is controlled by the muscles in the iris.

- The amount of light is reduced when the muscles of the iris contracts, reducing the opening of the pupil and making it smaller.

- The amount of light coming into the eye is increased when the muscles of the iris dialates or relaxes which in turn makes the opening of the pupil bigger.

Note: Dialate means expanding of any opening of the body, for example, dialation of the cervix during childbirth or the dilating pores of the skin.

The Ciliary Body

The **ciliary body** (SIL-ee-ehr-ee) is found in the choroid and it is a set of muscles and suspensory muscles used to improve the accuracy of focus of light rays on the retina. It does this by modifying the thickness of the lens.

The anterior segment of the eye is filled with the aqueous humor which is produced by the ciliary body. To focus on distance objects, the muscles modifies the lens making thicker while the muscles stretches the lens therby making it thinner in order to focus on distant objects.

The Cornea, Pupil, and Lens.

- The **cornea** (KOR-nee-ah) is the structure primarily focusing the light rays that enters the eye. It is the **transparent outer part of the eye** which covers the iris and pupil.

- The **pupil** is the part of the eye that allows light to enter the eye. It is the **black circular opening at the middle of the iris**.

- The **lens** is used to focus images on the retina and is a flexible, clear and curved structure. It is found behind the pupil and iris in a clear capsule.

Normal Action of the Eye

- **Accommodation** (ah-kom-oh-DAY-shun): This when the eyes adjusts it self to make it possible to see things at different distances. These adjustments could include changes in the shape of the lens, dialating or widening of the eye, movement of the eye, and contracting or narrowing of the eye.

- **Visual acuity** (ah-KYOU-ih-tee)**:** This is when the eye is able determine the shape and details of an object from a far distance. Acuity means clarity or sharpness.

- **Convergence** (kon-VER-jens)**:** This is and it is the inward movement of both eyes towards each other at the same time. This is done to maintain single binocular vision when an object comes closer.

- **Refraction:** This occurs due to the lens being able to bend light rays to enable them focus on the retina. Refraction is also referred to as **refractive power**.

- **Emmetropia** (em-eh-TROH-pee-ah)**:** It allows light rays focus properly on the retina due to the natural relationship between the shape of the eye and the refractive power of the eye. Emmetr means *proper measure* and -opia means *vision condition*

The Ears

The ears are the major organs used for hearing, and they do this by receiving sound impulses and sending them to the brain. The inner ear also aids in maintenance of balance.

Abbreviations Relating to the Ears

- Auris dexter (AD) is used to indicate the right ear.

- Auris sinister (AS)is used to indicate the left ear.

- Auris uterque or Auris unitas (AU) is used to indicate each ear or both ears respectively.

Auditory: audit - *sense of hearing* or *hearing,* and -ory - *pertaining to.* **Auditory** is (AW-dih-tor-ee and it is refers to the sense of hearing.

Acoustic (ah-KOOS-tick and it): acous - *sound* or *hearing.* It therefore pertains to hearing or sound.

The ear is seperated into three different parts: the inner ear, the middle ear and the outer ear as shown in figure 11.12.

The Inner Ear

- The inner ear is the part of the ear that houses the sensory receptors for both balance and hearing. **Labyrinth** (LAB-ih-rinth) is the term used to describe the structures present in the inner ear.

- The oval window is the structure that seperates the inner ear from the middle ear and also allows vibrations enter the inner ear. It is located at the underside of the stapes.

- The **cochlea** (KOCK-lee-ah), gotten from the Greek word *snail*, is a structure of the inner ear that is shaped like a snail. Its function is to convert sound vibrations into nerve impulses. The organ of corti, the acoustic nerves, the cochlear duct and the semicircular canal can all be found within the cochlea.

- The **organ of Corti** is that which gets vibrations from the **cochlear duct** and transmits them to the **auditory nerve fibers**. These fibers then relays the sound impulses to the auditory center of the brain's cerebral cortex and this is where the sound impulses are heard interpreted.

- There are three semicircular canals that house sensitive hair-like cells and the liquid endolymph. These hair-like cells help to maintain equilibrium as they bend response to the movements of the head and sets up impulses in nerve fibers.

- The **acoustic nerves** also known as the **cranial nerve** then relays this information to the brain and the then brain sends signals to muscles in all parts of the body to make sure equilibrium is maintained. The state of balance is used to define equilibrium.

The Middle Ear

- The **middle ear** is found between the inner ear and the outer ear. It sends sound across the space between the inner and outer ear.

- The **tympanic membrane** also known as the **eardrum** is located between the outer and middle ear. Tympanic (tim-PAN-ick,). When sound waves get to the eardrum, the membrane transfers the sound through vibration.

- The **mastoid process** is the temporal bone which contains the hollow air space that surrounds the middle ear.

The Auditory Ossicles

The **ossicles** (OSS-ih-kulz) are three small bones found in the middle ear. The function of these bones is to relay the sound waves from the eardrum to the inner ear through vibration. The names of these bones are gotten from Latin terms which describe their shapes. They are the:

- **Incus** (ING-kus) and it is known as the anvil

- **Stapes** (STAY-peez) and it is known as the stirrup

- **Malleus** (MAL-ee-us) and it is known as the hammer

The Eustachian Tubes

The eustachian tubes are also known as the **auditory tubes.** They are the narrow tubes leading from the middle ear to both the nasal cavity and the throat. The function of these tubes is to equalize the air pressure within the middle ear and that of the outside atmosphere. Eustachain (you-STAY-shun).

The Outer Ear

- The **pinna** also referred to outer ear or as the auricle and it is the external part of the ear. The pinna (PIN-nah) receives sound waves and sends them into the external auditory canal.

- The **external auditory canal** then sends these sound waves to the **tympanic membrane** or **eardrum** of the middle ear.

- **Cerumen** (seh-ROO-men) is known as the **earwax** which is secreted by the ceruminous glands that line the auditory canal. The earwax is a sticky yellow-brown substance which has protective functions such as traping dust, debris, some bacteria and small insects and preventing them from going into the middle ear.

Normal Action of the Ears

- **Air conduction:** This the way sound waves goes into the ear through the pinna and then moves down the external auditory canal until they get to the tympanic membrane. The **tympanic membrane** is found between the middle ear and the outer ear.

- **Bone conduction:** This happens when the eardrum vibrates and this makes the auditory ossicles of the middle ear to vibrate. Once the auditory ossicles begin to vibrate, the vibration of these bones begins to send sound waves through the middle ear down to the oval window of the inner ear.

- **Sensorineural conduction:** This happens when the sound vibration gets to the inner ear. The structures of the inner ear get these sound waves and transmit them to the auditory nerve before being transmitted to the brain. **Sensorineural** (sen-suh-ree-NOOR-al.

Chapter 15: Medical Specialties

General Medical Specialties Relating to Health and Disease

Specialists are medical practitioners who have been trained to care for the general wellbeing of their patients, examples of theses specialists are:

- **Family physicians** or **general practitioners** (GP) provide care for all patients, regardless of their ages.

- A **hospitalist** is an expert in the general medical care of patients on admission in the hospital.

- An **internist** is a physician who is an expert in the diagnosis and treatment of conditions and diseases of the internal organs and other systems in the body.

- A **gerontologist**, or **geriatrician** (jer-ee-ah-TRISH-un) is a physician who is an expert in caring for old people.

- A **pediatrician** (pee-dee-ah-TRISH-un) is a specialist who is an expert in the diagnosis, treatment and prevention of diseases and illnesses in infants and children. The field of discipline is called **pediatrics**.

Medical Specialties Related to the Skeletal System

- An **orthopedist** or an **orthopedic** (or-thoh-PEE-dick) surgeon is a medical practitioner who is an expert in the diagnosis of disorders and conditions related to the bones, muscle, and joints.

- A **rheumatologist** (roo-mah-TOL-oh-jist) is a medical practitioner who is an expert in diagnosis and treating arthritis and medical conditions like fibromyalgia, osteoporosis, and tendinitis that are due to inflammation in the connective tissues and joints.

- A **chiropractor** (KYE-roh-prack-tor) has a Doctor of Chiropractor degree and is an expert in the manual treatment of conditions that are as a result of spine displacement. Manual adjustment of the position of the bone is a form of manipulative treatment.

- A **podiatrist** (poh-DYE-ah-trist) is a Doctor of Podiatric Medicine (DPM) or a Doctor of Podiatry (DP) degree holder who is an expert in the diagnosis and treatment of diseases related to the foot. 'pod'- *foot*, '-iatrist'- *expert*.

- An **osteopath** (oss-tee-oh-PATH) has a Doctor of Osteopathy (DO) degree and deploys traditional methods of treatment and also uses manipulative methods in the treatment of medical conditions. This field of medical discipline is **osteopathy**. Osteopathy could also refer to bone diseases of any nature.

Medical Specialties Related to the Cardiovascular System

- A **hematologist** (hee-mah-TOL-oh-jist) is a medical practitioner who is an expert in the diagnosis of disorders, conditions, and abnormalities of the blood and tissues that form blood. 'hemat'- *blood*, '-ologist'- *expert*.

- A **cardiologist** (kar-dee-OL-oh-jist) is a medical practitioner who is trained in the diagnosis and treatment of disease, conditions, and abnormalities of the heart 'cardi'- *heart*, '-ologist'- *expert*.

- A **vascular surgeon** is a medical professional who have been trained to medically manage, diagnose, and perform surgical interventions for abnormalities relating to the blood vessel.

Medical Specialties Related to the Lymphatic And Immune Systems

The protection and maintenance of the health of the body is as a result of the collaborative efforts of the immune and lymphatic systems. Specialized structures, or common structures in both systems are responsible for ensuring optimal functioning of the body. Medical practitioners treat the conditions that might adversely affect the body if there is an improper functioning of these systems.

- A **lymphologist** (lim-FOL-oh-jist) is a medical expert who can proficiently diagnose and treat abnormalities of the lymphatic system. 'lymph'- *lymphatic system*, '-ologist'- *expert*.

- An **immunologist** (im-you-NOL-oh-jist) is a medical expert in the diagnosis and treatment of the immune system. 'immun'- *protected*, '-ologist'- *expert*.

- An **oncologist** (ong-KOL-oh-jist) is a medical professional who can expertly diagnose and treat malignant conditions like cancers and tumors.

'onc'- *tumor*, '-ologist'- *expert*.

- An **allergist** (AL-er-jist) is a medical professional who can expertly diagnose and treat conditions that are due to immunologic reactions like allergies.

Medical Specialties Related to the Respiratory System

- A **thoracic surgeon** is a medical professional who carries out surgical interventions on organs such as the lungs, esophagus, and heart that are within the chest and thoracic cavity

- A **pulmonologist** (pull-mah-NOL-oh-jist) is a medical practitioner who is an expert in the diagnosis and treatment of disorders and abnormalities relating to the respiratory system. 'pulmon'- *lung*, '-ologist'- *expert*.

- An **ENT physician** or an **otolaryngologist** (oh-toh-lar-in-GOL-oh-jist) is a medical practitioner who is a specialist in diagnosing and treating of diseases and abnormalities of the head and neck 'oto'- *ear*, 'laryngo'- *larynx*, '-ologist'- *expert*.

Medical Specialties Related to the Digestive System

- The field of medicine that is related to preventing and managing obesity and other related disorders is called **bariatrics**.

- A **dentist** is a Doctor of Medical Dentistry (DMD) or Doctor of Dental Surgery (DDS) degree holder who is an expert in the diagnosis and treatment of disorders and conditions of the tissues in the oral cavity, and the teeth.

- An **orthodontist** (or-thoh-DON-tist) is a dental professional who is an expert in the prevention and correction of malocclusion of the facial structures, including the teeth. 'orth'- *aligned*, 'odont'- *teeth*, '-ist'- *expert*.

- A **maxillofacial** (mack-sill-oh-FAY-shul) or **oral surgeon** is a medical expert who have been trained to perform surgical interventions on the jaws and face for effective treatment of diseases, repair of injuries, and correction of deformities.

- A **periodontist** (pehr-ee-oh-DON-tist) is a dental professional who is an expert in the prevention and treatment of diseases of the tissues around the teeth. 'peri'- *around*, 'odont'- *teeth*, '-ologist'- *expert*.

- A **gastroenterologist** (gas-troh-en-ter-OL-oh-jist) is a medical expert in the diagnosis and treatment of disorders and abnormalities of the internal organs including the intestines and stomach. 'gastro'- *stomach*, 'enter'- *small intestine*, '-ologist'- *expert*.

- A **proctologist** (prock-TOL-oh-jist) is a medical expert in the diagnosis and treatment of abnormalities of the colon, rectum, and anal region. 'proct'- anus and *rectum*, '-ologist'- *expert*.

Medical Specialties Related to the Urinary System

- An **urologist** (you-ROL-oh-jist) is a medical expert in the diagnosis and treatment of abnormalities ad diseases of the genitourinary system in men and the urinary system in women. 'Ur'- *urine*, '-ologist'- *expert*.

- **Nephrologists** (neh-FROL-oh-jist) are medical professional who are experts in the diagnosis and treatment of disorders and abnormalities relating to the kidneys. 'Nephr'- *kidney*, '-ologist'- *expert*.

Medical Specialties Related to the Nervous System

- An **anesthetist** (ah-NES-theh-tist) is an expert

in the administration of anesthesia, although is not a medical practitioner; for instance, a nurse anesthetist; 'an-' *absent*, 'esthesi'- *feeling*, '-ologist'- *expert*.

- A **neurosurgeon** is a medical practitioner have been trained to perform surgical intervention for disorders relating to the nervous system.

- A **neurologist** (new-ROL-oh-jist) is a medical practitioner who can expertly diagnose and treat disorders and conditions relating to the nervous system. 'Neur- *nerve*, '-ologist'- *expert*.

- An **anesthesiologist** (an-es-thee-zee-OL-oh-jist) is a medical expert in the administration of anesthetic medications prior to surgery, and in the course of surgery. 'An-' *absent*, 'esthesi'- *feeling*, '-ologist'- *expert*.

- A **psychologist** (sigh-KOL-oh-jist) holds a doctoral degree (PsyD or PhD); although is not a medical doctor (MD). A psychologist is an expert in the assessment and treatment of mental disorders and emotional issues. 'Psych'- *ind*, '-ologist'- *expert*.

- A **psychiatrist** (sigh-KYE-ah-trist) is a medical doctor (MD) who is an expert in the diagnosis and treatment of mental disorders, substance dependencies, and emotional issues. A psychiatrist has been trained in the expert prescription of drugs. 'Psych'- *mind*, '-iatrist'- *expert*.

Medical Specialties Related to the Eyes

- An **optician** (op-TISH-uhn) is a specialist who can expertly design and prescribe lenses for the correction of vision problems.

- An **optometrist** (op-TOM-eh-trist) has a doctor of optometry degree and is an eye care specialist who can expertly assess visual accuracy to evaluate the necessity of corrective lenses, and also diagnose conditions and disorders of the eye. 'Opt'- *vision*, '-metrist'- *a person that evaluates*.

- **Ophthalmologists** (ahf-thal-MOL-oh-jistz) are medical experts in the diagnosis and treatment of a broad range of disorders and abnormalities relating to the eye, including eye surgery and correction of vision. 'Ophthalm'- *eye*, '-ologist'- *expert*.

Medical Specialties Related to the Ears

- **Audiologists** (aw-dee-OL-oh-jist) are medical experts who accurately evaluate hearing and efficiently rehabilitates individuals with hearing problems. 'Audi'- *hearing*, '-ologist'- *expert*.

Medical Specialties Related to the Integumentary System

- A **plastic surgeon** is a medical expert who have been trained to perform surgical intervention to restore and realign various body parts. Here, plastic refers to the suffix '-plasty' which describes a procedure of **surgical repair**.

- Plastic surgeons who perform surgeries like face-lifts, liposuction, or breast augmentation, not for medical purposes, but for beautification purposes are called **cosmetic surgeons**.

- A **dermatologist** (der-mah-TOL-oh-jist) is a medical practitioner who is an expert in the diagnosis and treatment of skin conditions. 'Dermat'- *skin*, '-ologist'- *expert*.

Medical Specialties Related to the Endocrine System

- **Endocrinologists** (en-doh-krih-NOL-oh-jist) are medical practitioners who are experts in the diagnosis and management of endocrine glands abnormalities and disorders. 'Endocrin'- *producing inside*, '-ologist'- *expert*.

- A **certified diabetes educator** (CDE) is a professional who have been trained to educate people who have been diagnosed with diabetes on the most effective management of the disease.

Medical Specialties Related to the Male Reproductive System

- An **urologist** (you-ROL-oh-jist) is a medical expert in the diagnosis and treatment of abnormalities and diseases of the genitourinary system in men and the urinary system in women. **Genitourinary** means the genital bad urinary systems. 'Ur'- urine, '-ologist'- expert.

Medical Specialties Related to the Female Reproductive System and Childbirth

- **Obstetrician** (ob-steh-TRISH-un), or OB, are medical practitioners who are experts in the provision of skilled medical care to females in the course of pregnancy, delivery and the period following delivery. This field of discipline is called **obstetrics**.

- A **gynecologist** (guy-neh-KOL-oh-jist), or GYN is one who is a medical expert in the diagnosis and treatment of abnormalities and diseases relating to the female reproductive system. 'Gynec'- *women*, '-ologist'- *expert*.

- An **infertility expert** or a **fertility expert** provides medical care to infertile couples by helping them to diagnose and treat conditions that are related to conceptions and the maintenance of pregnancy.

- A **neonatologist** (nee-oh-nay-TOL-oh-jist) is a medical practitioner who is an expert in the diagnosis and treatment of medical conditions in the newborn 'Neo'- new, 'nat'- born, '-ologist'- expert.

- A **midwife** is a professional who have been trained to help pregnant women during childbirth. A certified nurse midwife (CNM) is a Registered Nurse who have been trained in obstetrics and gynecology, providing basic medical care during pregnancy and childbirth.

Conclusion

Learning medical terminologies might feel like a tedious chore. Infact, you could stay in on Fridays and Saturdays pouring so many medical terms just to find that you can only recollect about seven by Sunday or Monday. It's unavoidable if you're a medical student or professional.

In truth, it is very possible to know all the important terminologies by heart. All you need is the right approach. The most important of it being understanding word parts and word formations. This book is a proven resource to easily get by and become better than your peers. I'm sure that by now, you would have noticed that you learned most of the terms in here effortlessly. I imagine you having a rush of medical terms in your head right now. Don't worry, it's expected!

Beyond studying the words and learning word formations and word parts, you can remember complex terms by using use visual cues, practicing daily, using self-created acronyms, engaging friends and families, and using self-created word cards. Also, you could set apart a particular time during the day to learn each words, taking them chapter by chapter, and eliminating every distraction, if possible, use the library and other quiet places where you won't get interruptions. You should be able to complete this book many times in a year.

Only a few books are optimized with the main objective of teaching and helping you effectively memorize pronunciations, and this book tops the list. I can't conclude that you'll be perfect after reading it once, however, revisiting this book every once in a while will help you become better at memorizing them. The best way to learn these words is understanding what suits you most. Our brains will become wired to remember words fast these ways.

Do not forget, you can always revisit the different chapters in this book at different times.

www.ingramcontent.com/pod-product-compliance
Lightning Source LLC
Chambersburg PA
CBHW030503210326
41597CB00013B/773